A Guide to the Ergonomics of Manufacturing

A Guide to the Ergonomics of Manufacturing

Martin Helander

Linköping Institute of Technology, Sweden,
and State University of New York at Buffalo, USA

UK Taylor & Francis Ltd, 4 John St, London WC1N 2ET

USA Taylor & Francis Inc., 1900 Frost Road, Suite 101, Bristol, PA 19007

Copyright © Martin Helander, 1995

All rights reserved. No part of this publication may be reproduced, stored in a retrieval system, or transmitted, in any form or by any means, electronic, electrostatic, magnetic tape, mechanical, photocopying, recording or otherwise, without the prior permission of the copyright owner and the publisher.

British Library Cataloguing in Publication Data
A catalogue record for this book is available from the British Library.

ISBN 07484-0122-9

Library of Congress Cataloging-in-Publication Data is available

Cover design by Amanda Barragry.

Typesetting by Euroset, Alresford, Hampshire, England.

Printed in Great Britain by Burgess Science Press, Basingstoke, on paper which has a specified pH value on final paper manufacture of not less than 7.5 and is therefore 'acid free'.

To Mary

Contents

Foreword xiii

Chapter 1. Introduction 1
1.1 Brief History of Ergonomics 3
1.2 The Interdisciplinary Nature of Ergonomics 4
1.3 Ergonomics for Productivity, Safety, Health and Comfort 4

Chapter 2. Case Studies of Implementation of Ergonomics in Manufacturing 7
2.1 Ergonomic Improvements in Card Assembly 7
2.1.1 Design Improvements 8
2.1.1.1 Illumination Level 8
2.1.1.2 Special Lighting for Inspection 8
2.1.1.3 Job Rotation and Shift Overlap 9
2.1.1.4 Personal Music 9
2.1.1.5 Ergonomic Chairs 9
2.1.1.6 Operator Communication and Feedback 9
2.1.1.7 Materials Handling 9
2.1.1.8 Automation of Monotonous Jobs 9
2.1.1.9 Metric to Decimal Conversion 9
2.1.1.10 Housekeeping 9
2.1.1.11 Noise Reduction 10
2.1.1.12 Ergonomics Training 10
2.1.1.13 Continuous Flow Manufacturing 10
2.1.1.14 Evaluation of Protective Gloves 10
2.1.2 Specific Problems 10
2.1.3 Cost Efficiency of Improvements 10
2.2 Ergonomic Improvements in the Assembly of a Printer 12
2.2.1 Task Considerations 12
2.2.2 Workstation Ergonomics 13
2.2.3 Design of Tools and Controls 14
2.2.4 Discussion 14

Chapter 3. Anthropometry in Workstation Design 17
3.1 Measuring Human Dimensions 17
3.2 Definition of Anthropometric Measures 21
3.3 Using Anthropometric Measures for Industrial Design 24
3.4 Procedure for Anthropometric Design 25
3.4.1 Exercise: Designing a Microscope Workstation 27

Chapter 4. Physical Work and Heat Stress 29
4.1 Physical Workload and Energy Expenditure 29
4.1.1 Metabolism 29
4.1.2 Individual Differences 30
4.1.3 Metabolism During Work 30
4.1.3.1 Example: Calculation of Relative Workload 32

4.1.4 Measurement of Physical Workload 33
4.1.4.1 Example: Fatigue Due to Physical Workload 33
4.2 Heat Stress 33
4.2.1 Thermoregulation 33
4.2.2 Measurement of Heat Exposure 34
4.2.3 Wet Bulb Globe Temperature 34
4.2.4 Heat Stress Management 35
4.2.5 Comfort Climate 36
4.2.5.1 Example: Discussion of Heat Stress Measures 37

Chapter 5. Manual Lifting 39
5.1 Statistics of Back Injuries Associated with Lifting 39
5.2 A Biomechanical Model for Lifting 41
5.3 The So-called 'Correct Lifting Technique' 43
5.4 Guidelines and Standards for Lifting 46
5.4.1 1991 NIOSH Equation for Evaluation of Manual Lifting 46
5.4.1.1 Example: Loading Punch Press Stock 48
5.4.1.2 Example: Product Packaging 49
5.4.1.3 Lifting Index 50
5.4.2 Guidelines for the European Community 50
5.4.3 Guidelines for Manual Lifting in the UK 50
5.5 Materials Handling Aids 50
5.5.1 Materials Handling Devices 52
5.6 Recommended Reading 54

Chapter 6. Choice of Work Posture: Standing, Sitting, or Sit-Standing? 55
6.1 Examples of Work Posture 55
6.2 Identifying Poor Postures 58
6.2.1 Example: Sitting in India 58
6.3 Sitting, Standing or Sit-Standing 59
6.4 Hand Height and Determination of Table Height 60
6.4.1 Example 1 62
6.4.2 Example 2 62
6.4.3 Example 3 62
6.5 Work at Conveyors 63

Chapter 7. Repetitive Motion Injury 65
7.1 Carpal Tunnel Syndrome 65
7.2 Cubital Tunnel Syndrome 67
7.3 Tendonitis (or Tendinitis) 68
7.4 Tenosynovitis (or Tendosynovitis) 68
7.5 Thoracic Outlet Syndrome 68
7.6 Cause of Repetitive Motion Injury 69
7.7 Design Guidelines to Minimize Repetitive Motion Injury 71

Chapter 8. Hand Tool Design 73
8.1 Fitting the Task 73
8.2 Designing for the User 74
8.3 Prevention of Injuries 75
8.4 Segmental Vibration 77
8.5 Design Guidelines for Hand Tools 78

Chapter 9. Illumination at Work 79
9.1 Measurement of Illuminance and Luminance 79
9.2 Measurement of Contrast 80
9.2.1 Example: Contrast Requirements in Manufacturing 80
9.3 Use of a Photometer 81
9.4 Recommended Illumination Levels 82
9.5 The Ageing Eye 83
9.6 Use of Indirect (Reflected) Lighting 86
9.7 Cost Efficiency of Illumination 87
9.8 Special Purpose Lighting for Inspection and Quality Control 89

Chapter 10. Design of VDT Workstations 91
10.1 Sitting Work Posture 91
10.1.1 Viewing Angle 91
10.1.2 Thigh Clearance and Low-profile Keyboards 91
10.1.3 Chair Design 92
10.1.4 Supports for the Hands, Arms and Feet 93
10.1.5 Viewing Distance 94
10.2 Visual Fatigue 94
10.3 Effect of Radiation 95
10.4 Reducing Reflections and Glare on VDT Screens 96
10.4.1 Example: Calculating the Effect of a Neutral Density Filter on the Display Contrast Ratio 99

Chapter 11. Design of Controls 101
11.1 Appropriateness of Manual Controls for the Task 101
11.2 Computer Input Devices 103
11.3 Control Movements Stereotypes 104
11.3.1 Example: Controls for an Overhead Crane in Manufacturing 105
11.4 Control-Response Compatibility 106
11.5 Coding of Controls, Hand Tools, Part Bins and Parts 108
11.5.1 Coding by Location 108
11.5.2 Coding by Colour 108
11.5.3 Coding by Size 109
11.5.4 Coding by Shape 109
11.5.5 Coding by Labelling 110
11.5.6 Coding by Mode of Operation 110
11.5.7 Coding of Parts and Other Things Touched by the Hand 110
11.6 Emergency Controls 111
11.6.1 Example: Accidental Activation of Seat Ejection Controls in Aeroplanes 111
11.7 Organization of Items at a Workstation 111
11.8 Principles for the Design of Workstations 112
11.9 Recommended Reading 114

Chapter 12. Design of Symbols, Labels and Visual Displays 115
12.1 Symbols 115
12.1.1 Example: Standardization of Symbols 115
12.2 Labels and Written Signs 117
12.3 Warning Signs 117
12.3.1 Information Processing of Warning Signs 118
12.3.1.1 Information Overload 118
12.3.1.2 Attention and Active Processing 118
12.3.1.3 Comprehension and Agreement 119
12.3.1.4 Selecting and Performing a Response 120

Chapter 13. Development of Training Programmes and Skill Development 121
13.1 Establishing the Need for Training 121
13.2 Determining Training Content and Training Methods 122
13.3 The 'Why?', 'What?' and 'How?' of Training Development 123
13.4 Use of Task Analysis 124
13.5 Training in Manufacturing Skills 126
13.6 Part-task versus Whole-task Training 126
13.7 Use of Job Aids 127
13.7.1 Example: Remembering Error Codes 127
13.7.2 Example: Study of Job Aids 127
13.8 The Power Law of Practice 128
13.8.1 Example: Prediction of Future Assembly Time 129
13.9 Recommended Reading 130

Chapter 14. Noise 131
14.1 Measurement of Sound 131
14.1.1 Example: Calculation of Noise Dose 132
14.2 Noise Exposure and Hearing Loss 133
14.3 Hearing Protectors 134
14.4 Analysis and Reduction of Noise 134
14.4.1 Reduction of Noise in Manufacturing Plant 136
14.5 Effects of Noise on Performance 136
14.5.1 Broadbent and Poulton's Theories 138
14.5.2 Example: Discussion of Theories 139
14.6 Annoyance of Noise and Interference with Communication 139
14.6.1 Interference of Noise with Spoken Communication 140
14.6.1.1 Preferred Noise Criteria (PNC) Curves 140
14.6.1.2 Preferred Speech Interference Level (PSIL) 141

Chapter 15. Shift Work 143
15.1 Example: How Not to Schedule Shift Work 144
15.2 Circadian Rhythms 144
15.3 Problems with Shift Work 145
15.4 Effects on Performance and Productivity 146
15.5 Improving Shift Work 147
15.5.1 Type of Work 147
15.5.2 Shift Work Schedules 148
15.5.3 Selecting Individuals for Shift Work 149
15.6 Recommended Reading 150

Chapter 16. Whole Body Vibration 151
16.1 Sources of Vibration Discomfort 151

Chapter 17. Design for Manufacturing Assembly 155
17.1 The Desire to Automate 155
17.2 What to Do and What to Avoid in Product Design 157
17.2.1 Using a Base Part as the Product Foundation and Fixture 157
17.2.2 Minimizing the Number of Components and Parts 158
17.2.3 Facilitating Handling of Parts 159
17.2.4 Facilitating Orientation of Parts 160
17.2.5 Facilitating Assembly 161
17.2.6 Consideration of Stability and Durability 161
17.3 Designing Automation using Boothroyd's Principles 163
17.4 MTM Analysis of an Assembly Process 163

17.5 Human Factors Principles in Design for Assembly 165
17.5.1 Example: Design for Job Satisfaction 169

Chapter 18. Design for Maintainability 171
18.1 Ease of Fault Identification 171
18.2 Design for Testability and Troubleshooting 173
18.3 Design for Accessibility 174
18.4 Design for Ease of Manipulation 174
18.5 Summary 174

Chapter 19. Machine and Robot Safety 177
19.1 Safety Devices 177
19.1.1 Physical Barriers 177
19.1.2 Photoelectric Beams 178
19.1.3 Pressure-sensitive Mats 178
19.1.4 Infrared Sensors 179
19.1.5 Cameras and Image Processing 179
19.1.6 Ultrasound (Sonar) 179
19.1.7 Capacitive Sensors 179
19.2 Example: Case Study of Robot Safety at IBM Corporation/Lexmark 180
19.3 Recommended Reading 180

References 181

Appendix
The Use of an Ergonomics Checklist in Manufacturing 197

Index 205

Foreword

During the past 10 years I have taught a course to industry on ergonomics in manufacturing. The audience typically included engineers, company nurses, medical doctors, managers and workers. This book builds on that course and the information presented is what filtered through after critique and suggestions for modifications by several thousand students. The book is intended for engineers and students of engineering who design manufacturing systems and workstations. The text is useful to human factors/ergonomics experts who want to understand manufacturing applications of ergonomics. Physiotherapists, occupational nurses and medical doctors who take an interest in manufacturing will also find relevant information.

I had the pleasure of collaborating with several individuals in industry. My thanks go first to George Burri who initiated ergonomics training at IBM Corporation. I have particularly enjoyed the collaboration with Daniel Bentivogli, Edward Grossmith, Kal Kawar and Michael Heffernan. I learned much from their experience and enthusiasm for implementing ergonomics in the real world.

I started writing this text when I was at SUNY Buffalo, USA, and finished it at Linköping Institute of Technology in Sweden.

I am grateful to my students Richard Adorante, Jennifer Buckshaw, Michael Edwards, Ahmed Mohamood, Margaret Price, John Satterlee and Kelvin Si who read the text and made many suggestions for improvement.

For the preparation of the manuscript I was fortunate to work with Patricia Brock. Her efforts, dedication and enthusiasm were instrumental in the completion of this book.

Chapter 1

Introduction

The word *ergonomics* comes from the Greek *ergo* (work) and *nomos* (law). It was used for the first time by Wojciech Jastrzebowski in a Polish newspaper in 1857 (Karwowski, 1991). In the USA, *human factors engineering* or *human factors* have been close synonyms. European 'ergonomics' has its roots in work physiology, biomechanics and workstation design. 'Human factors', on the other hand, has its origin in experimental psychology and the focus is on human performance and systems design (Chapanis, 1971).

There are common problems in the workplace where it is necessary to take a broad approach. Despite the differences between human factors and ergonomics in the type of knowledge and design philosophy, the two approaches are coming closer. For example, the introduction of computers in the workplace presents a variety of design problems (see Table 1.1). We can illustrate the problem as shown in Figure 1.1. Here a human operator is perceiving information on a display. The information is then interpreted and an appropriate action is selected. The action is executed manually as a control input, which in turn effects the information status on the display.

The environment may also affect the human operator. Here it would be appropriate to analyse factors that are external to the task and yet may have a great effect on performance and job satisfaction, for example:

- Noise and vibration.
- Heat and cold.
- Work–rest cycle.
- Organizational factors.

To effectively solve a problem related to VDT workplaces, an ergonomist must be able to recognize and analyse a variety of problems and suggest design solutions. This leads to our first maxim: *the primary purpose of ergonomics is design*.

Table 1.1 Design problems arising from the introduction of computers in the workplace

Problem	Knowledge required to solve problem
Work posture	Biomechanics
Keying	Biomechanics
Size of screen characters	Perception, vision research
Layout of screen information	Cognitive psychology, cognitive science
Designing new system	Systems design and cybernetics
Environmental factors	Noise, heat stress, cold stress

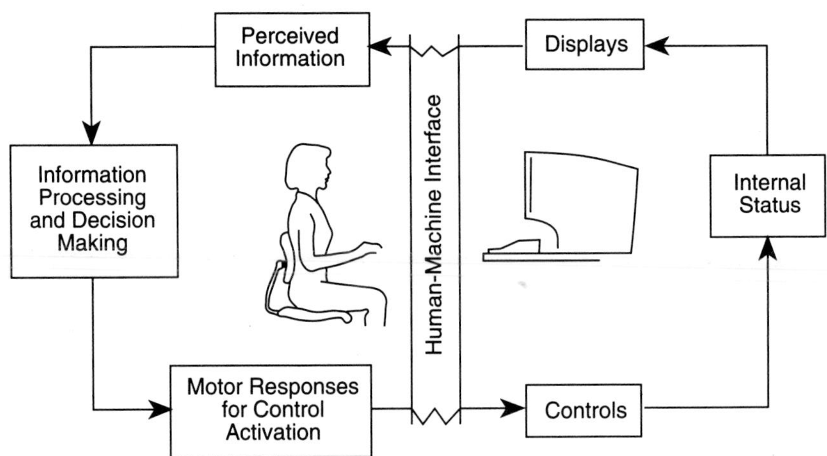

Figure 1.1 Analysis of the human–machine interface requires interdisciplinary knowledge of biomechanics, cognitive psychology and systems design methodology

The existing situation must, therefore, first be analysed, design solutions must be generated and these design solutions must be analysed. The design work can be described using a control loop, as shown in Figure 1.2.

It follows from Figure 1.1 that interdisciplinary knowledge is required: (1) to formulate systems goals; (2) to understand the functional requirements; (3) to design a new system; (4) to analyse the system; and (5) to implement the system. From the feedback loops shown in Figure 1.2 it also follows that design is a never-ending activity. There are always opportunities for improvements or modifications.

A common scenario for the work of an ergonomist could be the following: Imagine that the system shown in Figure 1.1 could be redesigned. Maybe there could be two displays, or perhaps part of the human information processing could be done by a computer, or maybe the manual input to the computer system could be made by computer voice recognition. In the redesign of the system the ergonomist would have to consider many constraints. There will be constraints in allocating tasks (who does what), economic constraints, company constraints, and sometimes labour union constraints. The ergonomist will obtain information from those who use the system or

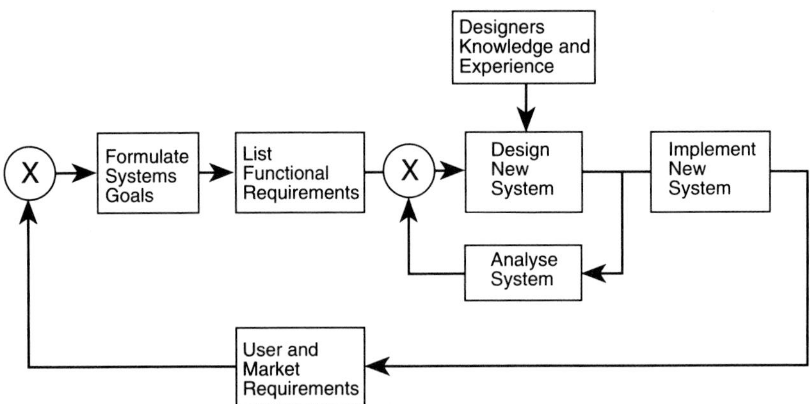

Figure 1.2 Procedure for design and redesign of a system

from another similar system. It will be necessary to consult textbooks and scientific articles, and in the end it may be necessary to evaluate several design options by using rapid prototyping or by performing an experiment with users as test subjects. This scenario leads to our second maxim: *a systematic, interdisciplinary approach is necessary in system design and analysis.*

1.1 Brief History of Ergonomics

In the USA, human factors emerged as a discipline after World War II. There were many problems encountered when using sophisticated war equipment such as aeroplanes, radar and sonar stations, and tanks. Sometimes these problems caused human errors with grave consequences. For example, during the Korean War, more pilots were killed during training than in war activities (Nichols, 1976). This finding focused the interest on the design of controls and displays in aircraft. How could information be better displayed, and how could controls be redesigned and integrated with the task so that they were easier to handle? Many improvements were implemented, such as a pilot's joystick which combined several control functions and made it easier to handle the aeroplane and auxillary combat functions (Wiener and Nagel, 1988). As a result of these improvements and new pilot training programmes, the number of fatalities in pilot training decreased to a fraction (5%) of what they had been previously. Ever since, most of the research in human factors in the USA has been sponsored by the Department of Defense. Consequently, the information available in textbooks on human factors is heavily influenced by military rather than civilian applications of ergonomics.

Some federal agencies have sponsored research on civilian applications: the Federal Highway Administration (design of highways and road signs), NASA (human capabilities and limitations in space, design of space stations), the National Highway Traffic Safety Administration (design of cars, including crash worthiness; effects of drugs and alcohol on driving), the Department of the Interior (ergonomics in underground mining), the National Bureau of Standards (safe design of consumer products), the National Institute of Occupational Safety and Health (ergonomic injuries at work, industrial safety, work stress), the Nuclear Regulatory Commission (design requirements for nuclear power plants), and the Federal Aviation Administration (aviation safety).

In the USA, applications in manufacturing are fairly recent. Eastman-Kodak in Rochester, New York, was probably the first company to implement a substantial programme around 1965. Their approach has been well documented in two excellent books (Eastman Kodak Company, 1983, 1986). At IBM Corporation, interest in manufacturing ergonomics started around 1980. At that time IBM had many human factors experts, but most of them worked on consumer product design. Currently they have turned their interest to computers and software systems. Most of the manufacturing ergonomics has been undertaken by industrial engineers and company nurses. Ergonomics is also discussed in 'quality groups', which comprise a mix of engineers and operators (Helander and Burri, 1994).

In Europe, ergonomics has had a different history. The discipline is particularly well established in the UK, France, Germany, Holland, Italy, and the Scandinavian countries. In the former USSR, just as in the USA, the interest was focused primarily on Department of

Defence activities. There have been few applications on the industrial side, but interest is quickly growing.

In many European countries, labour unions have taken an active interest in promoting ergonomics as being important for safety, health, comfort and convenience. The labour unions are particularly strong in the Scandinavian countries and in Germany, where they can often dictate what type of production equipment is purchased.

One may argue that ergonomics is nothing new. Even during the Stone Age individuals were designing hand tools to fit the user and the task (Drillis, 1963). During the Industrial Revolution there were efforts to apply the concepts of a 'human centred design' to tools such as the spinning-jenny and the spinning-mule. The concern was to allocate interesting tasks to the human operator, but let the machine do repetitive tasks (Rosenbrock, 1983). At the beginning of the 20th century, Frederick Taylor introduced the 'scientific' study of work. This was followed by Frank and Lillian Gilbreth who developed the time-and-motion study and the concept of dividing ordinary jobs into several small micro-elements called 'therbligs' (Konz, 1990). Today there are sometimes objections against Taylorism, which has been seen as a tool for exploiting workers. Nonetheless, these methods are useful for measuring and predicting work activities. The time-and-motion study is a valuable tool if used for the right purpose!

It was not until the 1950s that ergonomics became an independent discipline. In the UK, the Ergonomics Research Society was formed in 1950. In the USA, the Human Factor Society was established in 1957. In 1961 the first meeting of the International Ergonomics Association was held in Stockholm, Sweden (Chapanis, 1990). Today, this umbrella association represents about 15 000 ergonomists in 40 countries.

1.2 The Interdisciplinary Nature of Ergonomics

Ergonomists come from a variety of professional fields. This mixed background is well demonstrated by the membership of professional societies which typically consists of engineers, psychologists, and individuals from the medical profession.

To successfully implement ergonomics in manufacturing design and planning, it is often an advantage to be an engineer. Psychologists, medical doctors and industrial nurses can certainly diagnose many ergonomics problems, but sometimes have an insufficient technical background to suggest how a technical system can be redesigned. Engineers with a background in ergonomics are ideal, as they can analyse different design alternatives for machinery and processes, make trade-offs in the selection of equipment, and arrive at a better solution. Ergonomics is often implemented by work groups where the members have expertise in different areas. Groups composed of workers, engineers, managers and nurses can propose new design solutions. The establishment of such groups is typical of the complex decision-making found in modern manufacturing.

1.3 Ergonomics for Productivity, Safety, Health and Comfort

In many industries ergonomics is implemented primarily as a means of reducing high injury rates and high insurance premiums. In the USA, a worker's compensation premiums often amount to 15% of the salary. This is because there are many back injuries due to materials handling and injuries to the joints in the arms, shoulders and neck due to poor work posture.

During the past 5 years many injuries due to cumulative trauma disorders, carpal tunnel syndrome and tenosynovitis have been reported. At the same time, the number of back injuries remains high, and is still the main cause of industrial injury. It is estimated that the actual cost of musculoskeletal disease in the UK exceeds £25 billion. The reporting of injuries is affected not only by the actual injury, but also by psychological and sociological factors. A study by Hadler (1989) compared disabling back injuries in France, Switzerland and The Netherlands. He observed that not only are the legislative programmes in the three countries different, but the pattern of reported injuries is different. The conclusion was that, in addition to actual injuries, there are several psychological, attitudinal and ethical factors which determine what is reported as an accident or injury and what remains unreported. Individuals will sometimes report particular symptoms because they are 'recognized' by the country's legislation or by society. Different countries might pool injuries under different names. One interesting difference is between VDT operators in the Scandinavian countries and the USA. In the USA there is a prevalence of injuries due to cumulative trauma disorder and tenosynovitis of the hand and of the wrist. These types of injuries are more rare in the Scandinavian countries, where operators complain more about pain in the neck and the shoulder. Certainly there must be a connection between the two, but the prevalent ethic of one country is different from that in the other.

While the reduction of injuries and improved health of workers is a very important reason for implementing ergonomics, it is a fairly negative one. Management is forced to implement ergonomic measures to reduce the injury rate. The author is concerned that this 'negative' message will dominate, so that industry leaders will ignore what could be a much more important driving factor for ergonomics, namely increases in productivity. Ergonomic improvements in workstations, industrial processes, and product design can be undertaken from the point of view of productivity, and there can be tremendous gains. Management is often unaware of poor working conditions, and what types of improvement could improve productivity. Workers in plants and in offices usually adapt to the poor conditions – but the cost is increased production time, lower quality of production and, of course, increased injury rate. The two case studies in Chapter 2 illustrate the potential of ergonomics to improve productivity.

Ergonomics is also highly related to industrial safety. If workers can perceive hazards, if there are relevant warning signs, if controls are easy to use, if work postures are acceptable, if noise and other environmental stressors are reduced, if there is collaboration between workers and management based on mutual understandings, and if there is good housekeeping, then safety will improve. Ergonomics measures regarding safety are somewhat different from the conventional, somewhat mechanistic approach often taken in industrial safety. Ergonomics can improve safety through worker's attitudes, perception, decision-making, and risk-taking behaviour.

Figure 1.3 summarizes how an ergonomics systems analysis can be undertaken with at least three different objectives in mind: (1) ergonomics, (2) production, and (3) quality of manufacturing.

In the design of any complex system it becomes necessary to apply many criteria simultaneously. All these criteria must be at least partially satisfied or, to use Simon's (1969) terminology, multiple criteria must

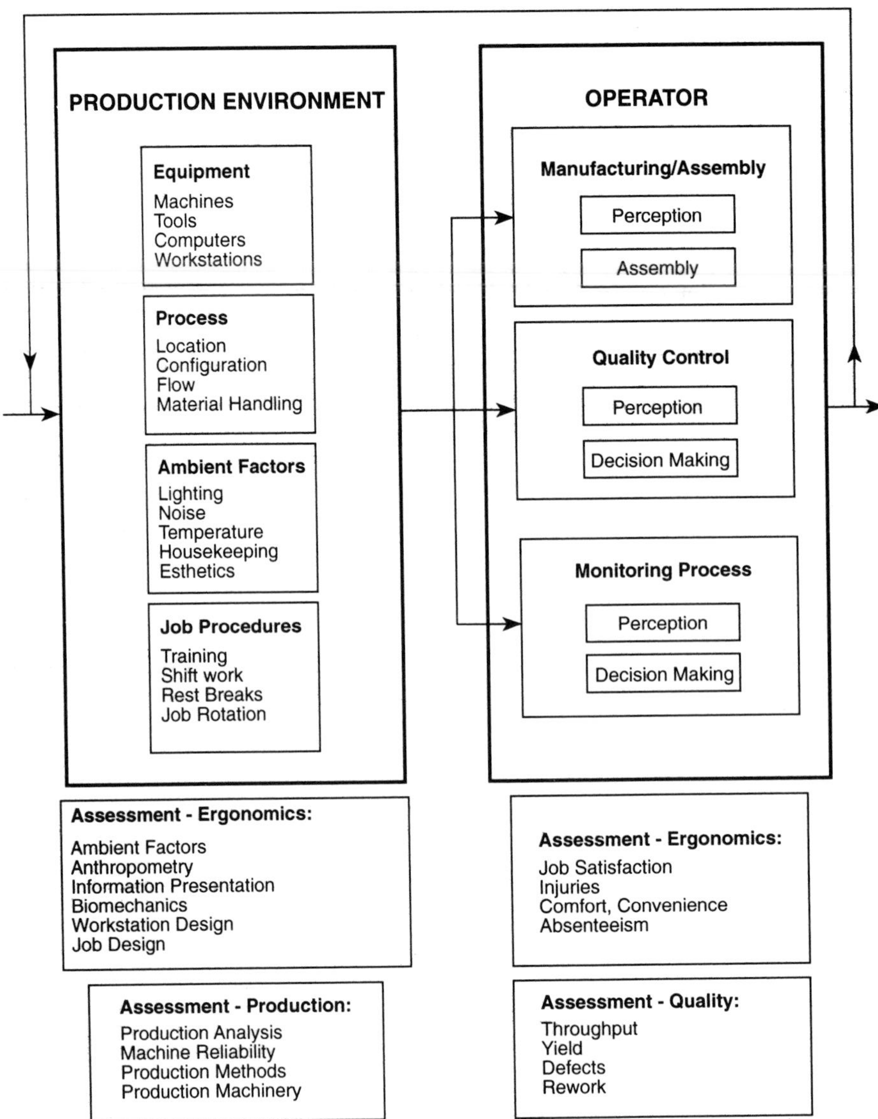

Figure1.3 A production environment / operator system. There are three broad criteria for assessment: ergonomics, production, and quality

be 'satisficed'. In other words, one cannot accept a manufacturing situation where either the production process, ergonomics, or quality of manufacturing are substandard. All assessment criteria must be at a certain minimum level to be acceptable.

The two case studies in Chapter 2 illustrate how ergonomic improvements can be implemented in manufacturing. The ergonomic improvements improved all aspects of system performance. There were no (obvious) conflicts between ergonomics and productivity – a win-win situation, as they say.

Chapter 2

Case Studies of Implementation of Ergonomics in Manufacturing

In this chapter two case studies are presented. Both studies illustrate how traditional ergonomics improvements were implemented. The first study was in an electronics plant. In this case the economic benefits from improved productivity far exceeded any other benefits accrued from worker comfort or reduced injury rate. This implementation of ergonomics was very successful, and the quality in manufacturing improved significantly.

The second case study took place in a plant that manufactured printers. Both manual labour and automation were used for assembly, and this study illustrates the difficulty in allocating tasks between automation and human operators. The surprising outcome was that, after a few years, the very expensive automated manufacturing was discarded. The work is now done manually.

2.1 Ergonomic Improvements in Card Assembly

At IBM in Austin, Texas, printed circuit boards for computers were manufactured. The boards consisted of multiple layers of copper sheet and fibre glass with etched circuitry. Holes were drilled through the circuit board for insertion of components. Much of the component insertion was automated. However, there were many tasks which could not be automated, including quality control and inspection of component parts and finished products.

One important measure of quality in the manufacturing of boards is the percentage production yield. In this case, plant management had observed that the yield was consistently 5–10% below target (Burri and Helander, 1991a). Most of the quality problems were described as 'internal', which means that there were defects inside the circuit board, which could have occurred at several different locations in the manufacturing process. It was therefore difficult to isolate problems as occurring in any specific department. In this study, we focused on one department called 'Core Circuitize'. This was located just prior to the determination of the percentage yield, about half-way through the manufacturing process. Altogether 132 individuals, mostly operators, worked at this location, consisting of 59 workstations.

To collect information and evaluate the manufacturing scenario, information was collected from five different sources:

1. Discussion with management.
2. Plant walk-through, inspection, and note taking.
3. Discussion with operators.
4. Discussion with first-line supervisors.
5. Field measurements of illumination, noise, and workstation design.

Through these discussions and measurements, data were gathered on the effectiveness of the operation. This provided the basis for a comprehensive assessment of both the system and the individual tasks, which revealed significant opportunities for improvements.

Most of the 59 workstations were different and it is not meaningful to summarize the data collected; we instead focus on the recommendations. Based on the information, we identified 14 design improvements (see Table 2.1). Some of these were conventional ergonomic measures, and some required redesign of the manufacturing process.

2.1.1 Design Improvements

2.1.1.1 Illumination Level

The improved illumination turned out to be the most important of all the measures. Several operators were performing a relatively simple task by placing circuit boards into machines for automatic insertion of components (so-called 'card-stacking machines'). The managers thought of the operators as supervisors of the automatic machines. However, interviews with the operators disclosed that they regarded themselves more as quality inspectors than as machine tenders. They would inspect cards and components that were placed in the machine and they inspected the finished product as it was removed from the machine. One of the most critical aspects of this task was to inspect the magazines containing the electronic components that were put into the card stacking machines. A common problem was that components were turned in the wrong direction in the magazines.

The average ambient illumination level was about 500 lux, which is inadequate for inspection work. In some areas the illumination was as low as 120 lux. It was decided to increase the illumination to 1000 lux throughout the department. This was achieved by installing fluorescent light tubes, switching on lights that had been turned off for energy-conservation reasons, and lowering light fixtures from high bay ceilings to a location closer to the workstations.

2.1.1.2 Special Lighting for Inspection

In addition to the above measures, some polarized lights were installed to make it easier to see imperfections and quality defects. Many examples of special illumination systems for inspection are presented in Chapter 9.

Table 2.1 Ergonomic improvements at the IBM plant in Austin, Texas

1. Uniform illumination level at 1000 lux
2. Installation of special lighting for inspection
3. Job rotation to avoid monotony
4. Personal music was distracting and was discontinued
5. Ergonomic chairs certified for clean rooms
6. Improved communication
7. Materials-handling guidelines
8. Automation of monotonous jobs
9. Metric to decimal conversion charts
10. Housekeeping improved
11. Noise reduction
12. Ergonomics training
13. Continuous flow manufacturing
14. Use of protective gloves

2.1.1.3 Job Rotation and Shift Overlap

Visual inspection is often monotonous, and there are problems in sustaining the attention throughout an entire work shift. To break the monotony, job rotation was incorporated so that operators could split their time between two jobs (Grandjean, 1985). Existing rest-break patterns were evaluated, but it did not seem necessary to increase the length of the rest break. The time overlap between shifts was reduced from 30 to 12 minutes. The shift overlap is generally used to transfer information between shift crews on the status of machines and processes. However, the existing overlap of 30 minutes was excessive.

2.1.1.4 Personal Music

An experiment was performed to introduce personal music in the workplace. However, the music was distracting to the work and it was therefore discontinued.

2.1.1.5 Ergonomic Chairs

New ergonomic chairs were provided to increase comfort. This also seemed to increase productivity, for operators could remain seated during inspection. The chairs were manufactured to be used in a clean-room environment. There were several adjustabilities including seat height, back-rest angle and seat-pan angle. For some operators, sit/stand types of chair were also provided for occasional use.

2.1.1.6 Operator Communication and Feedback

In order to enhance verbal communication and feedback between operators, openings were installed between some of the workstations. The openings improved communication significantly, particularly with respect to quality control (Bailey, 1982).

2.1.1.7 Materials Handling

Guidelines were established to limit the heights of storage racks. For example, the lowest shelf in the storage racks was removed. This made it impossible for anything to be put on the lower shelf, which reduced the amount of bending and back injuries. In addition, a guideline for a maximum weight of parts was established.

2.1.1.8 Automation of Monotonous Jobs

Some operations were converted from manual work to robot/automation. One of the jobs involved a task where a protective Mylar peel tape was removed from a board. This was a highly monotonous and repetitive task and did not provide any job satisfaction. The operator now supervises the robot, and in addition performs several other tasks, which provides for a more varied and interesting job.

2.1.1.9 Metric to Decimal Conversion

The conversion between metric and decimal measurements was confusing to several operators, and a conversion chart was provided for each workstation.

2.1.1.10 Housekeeping

Through collaboration with management, an example of good housekeeping was set up in part of the plant. The area was cleaned up and organized. This inspired operators in other areas as well, and housekeeping improved. As part of the housekeeping effort, the manufacturing facility was converted to a 10 000-type clean-room facility. Clean-room clothing and smocks were evaluated and their use recommended.

2.1.1.11 Noise Reduction

The noise levels were well within the 85 dBA stipulated by NIOSH. However, to enhance verbal communication, sound insulation covers were installed for several processes. The ambient noise level at the workstations was reduced from about 75 to 60 dBA, which greatly enhanced verbal communication between operators as well as their comfort.

2.1.1.12 Ergonomics Training

An ergonomics training and awareness programme was provided during departmental meetings. This was a 4-hour programme and addressed a variety of problems, very much similar to the contents of this text.

2.1.1.13 Continuous Flow Manufacturing

Continuous flow manufacturing was implemented at several locations in the plant. The main purpose was not to enhance ergonomics, but to reduce the amount of space required for manufacturing. There were important side benefits in that the distance between nearby operators decreased and verbal communication became possible.

2.1.1.14 Evaluation of Protective Gloves

For several of the operations, anti-laceration gloves were used to protect the operators from sharp edges and the corners of the boards. However, some types of glove reduced tactile sensation so that it was difficult to manipulate components. Several different gloves were tested, and the type of glove selected maximized tactile sensation.

2.1.2 Specific Problems

In addition to the general problems there were specific problems at several workstations. For the 'drills' operation, operators had to bend over the machine to replace drill bits that were used to drill holes in cards (see Figure 2.1). On the old machine, operators had to bend very carefully otherwise the drill bits would stick in their stomachs. In the new machines the drill bits were relocated, reducing the operator's reaching distance and improving the work posture. Using time-and-motion studies, time savings were calculated to be 1.5 minutes per set-up. This translated into a yearly saving of $270 000.

2.1.3 Cost Efficiency of Improvements

Based on our experience gained in previous ergonomic field studies, we had projected a 20% improvement in process yield, a 25% improvement in operator productivity, and a 20% reduction in injuries. Actual improvements were close to our predictions and resulted in a cost reduction of $7 375 000 (see Table 2.2). The cost of materials for ergonomic improvements (such as improved illumination) was $66 400. The labour costs for the implementation were about $120 000. The benefit/cost ratio for these improvements was approximately 40:1 for the first year or, phrased differently, the payback time was about 1 week (Helander and Burri, 1994).

Reductions in injury costs were fairly minor compared with the improvements in productivity and yield, demonstrating that improvements in productivity are sometimes extraordinary. We can conclude that ergonomics is important in improving the quality in production. Management was impressed by the results and hired two ergonomists with an industrial engineering background to continue the improvements.

There were also improvements in operator comfort, convenience and job satisfaction. Informal interviews were held with a large number of operators and with management. They showed that there were no

Figure 2.1 The location of the drill bits forced excessive reaching and required great caution. After the modification, drill bits were located close to drills, thus reducing reach

Table 2.2 Projected and actual improvements

	Improvement (%)		Cost reduction ($)	
	Projected	Actual	Projected	Actual
Yield improvement	20	18	2 268 800	2 094 000
Operator productivity	25	23	5 647 500	5 213 000
Injury reduction	20	19	73 400	68 000
Total			7 987 700	7 375 000

negative effects of the new system. Operators generally appreciated what had been done and were happy with the new system. These types of improvement are more tangible and difficult to quantify in terms of cost savings than are improvements in productivity and safety.

From a scientific point of view, this study is very unsatisfactory, for there was no control group for comparison. Such is often the case in industry. It is usually very difficult to find an identical control group, and for 'core circuitize' with 200 operators it was impossible. We are therefore restricted to selective evaluation methods without statistical significance testing. In our study, 26 managers and engineers were interviewed. They agreed that approximately half of the savings could be attributed to ergonomics, while the remaining half was attributed to other improvements such as the continuous flow manufacturing. Management were extremely positive about the ergonomic improvements, and particularly the increased illumination levels for visual inspection.

This case study also demonstrates that ergonomic improvements cannot be undertaken in isolation of the manufacturing process. There must be a clear understanding of technological alternatives for improving productivity and how ergonomics is affected by the choice of technical system, process layout and equipment.

2.2 Ergonomic Improvements in the Assembly of a Printer

The Proprinter manufacturing line at IBM in Charlotte has been well documented. The design of the printer presents an interesting case study of design for automation (DFA) concepts (Boothroyd and Dewhurst, 1983). The printer was redesigned several times in order to reduce the number of parts and to make it possible to assemble the entire printer by using automation. For example, only one type of screw fastener was used.

Altogether there were 18 workstations (4 manual and 14 automated). Total automation – an engineer's dream – turned out to be impractical. Rather, the production line was designed with flexibility in mind, so that workstations could be switched between manual or automatic. In this assembly there was no materials handling such as carrying or pushing. All parts were automatically delivered to the workstations.

The task allocation between automation and manual labour was based on considerations of what a robot (or automaton) could do, given the constraints of the manufacturing tasks. Manual tasks were typically tasks that were 'left-over', since they were either difficult to automate (such as cable routing), too costly to automate, or would take too much time to automate. The time to set up the automation is not much of a restriction, but the delivery time of automated production machinery can be excessive.

From the ergonomics perspective, the challenge was to design a product that was compatible with the requirements of the workforce. Most of the engineers at this location were familiar with ergonomics design principles and had participated in ergonomics training offered by the company. Several ergonomic design measures were taken and these are summarized below in Tables 2.3–2.5.

2.2.1 Task Considerations

Several measures were taken to improve task design (Table 2.3). To balance the assembly line, a 30-second cycle time was used. However, the implementation team increased the task duration time for the human operator from 30 seconds to 2.5 minutes. This was accomplished by integrating several manual tasks. Such longer, and

Table 2.3 Task design measures taken to improve ergonomics and job satisfaction

1. Increased assembly cycle time from 30 seconds to 2.5 minutes
2. Conveyor storage to reduce operator pacing by assembly line
3. Flexible schedule for rest breaks
4. Job rotation – two jobs per day
5. Buddy system for robotic safety
6. Physical proximity to enhance verbal community
7. Performance feedback to operators
8. Career path established for operators

more comprehensive tasks are considered more satisfying, since there is both less boredom and less fatigue due to the one-sided strain.

Job rotation is a technique that has been widely used to increase the competence of workers and to reduce monotony. In this facility, operators changed workstation twice per day going 'upstream'. Thereby the operators did not have to check on their own work from the previous operation, and they could feel free to report quality defects.

Several operators maintained robotics workstations. For maintenance and repair it is necessary to work close to the robot. To improve safety there was a buddy system with two individuals, one doing the maintenance and the other watching the robot. If something went wrong, it was possible to shut off the system very quickly.

Physical proximity is essential to promote team work between operators, engineers and management. To facilitate communication, the line manager's office was moved from a remote area to the manufacturing line. Support engineers' offices were located close by.

Performance feedback is important for all production. In this case task performance data were summarized and made available to the operator for evaluation, recognition and possible corrective action.

For this manufacturing operation several different job titles were established: assemblers, set-up operators, quality and process auditors, and automated equipment operators. This formalization of the job titles was intended to facilitate a career path and discussions between management and operators. It could also reinforce teamwork.

2.2.2 Workstation Ergonomics

Table 2.4 summarizes the measures that were taken to improve the workstation.

The worksurface height was made easily adjustable by using a motor drive. Thereby both the small and large operators could fit at the workstations.

Vacuum suction cups were attached to the worksurface and used to hold the kit in a comfortable position for assembly.

The roller ball transfers were used to facilitate the movement from the conveyor line to the workstation. Thereby the operator could push and pull the unit into the workstation and did not have to lift it.

Ergonomic chairs were acquired. The seat height and seat-back angle were adjustable. Thus operators could either sit or stand in the workplace.

Each workstation was equipped with tack boards for personal items, such as photographs. A lockable drawer served as personal storage for private items.

Table 2.4 Ergonomic design of the workstation

1. Work-table height electrically adjustable
2. Vacuum suction cups used to hold the unit in different positions
3. Roller ball transfer from conveyor line to workstation
4. Ergonomic chairs for sitting or standing work posture
5. Tack boards for personalization of workstation
6. Lockable personal storage
7. Use of low-noise equipment and enclosure of noise sources
8. 1000 lux ambient illumination and optional task illumination
9. Heat and ventilation optimized for comfort

Many researchers have reported the adverse affect of noise on job satisfaction. With the high degree of mechanization in the area, there was a great risk that conveyors, pneumatic devices and motors would create too much noise. Consequently, special attention was given to procure low-noise equipment and acoustic covers to enclose the sound at the noise source. For one operation (frame serialization), the noise was practically eliminated by using laser etching rather than stamping.

One very important factor is to provide adequate lighting at the workstation. Precision tasks typically need at least 1000 lux. In addition, moveable task lights were provided for each workstation.

Heating and ventilation requirements were established in accordance with IBM facility practices, to ensure comfortable climatic conditions.

2.2.3 Design of Tools and Controls

Measures were taken to procure ergonomic handtools, controls, and displays (Table 2.5).

The power tools used in the operation were lightweight and more delicate in appearance than common heavy-duty power tools typically used in manufacturing. This increased the efficiency and comfort of the operator (Hasselquist, 1981).

Some of the controls on the control panel were removed as they were unnecessary for the operation. This reduced the number of possible actions, which simplified the task. In addition, schematic lines were drawn to identify control sequences. Logical groupings of controls were enhanced by using contrasting colours.

One common problem in maintenance is the lack of adequate lighting. To simplify maintenance, permanent light fixtures were installed inside equipment panels.

2.2.4 Discussion

The ergonomic design took place during the preliminary system design phase. Because these features were implemented at a very early stage, they were inexpensive.

All the parties involved (operators, managers and engineers) were very enthusiastic about the ergonomic design. The operators appreciated the design of the workstation, including the flexibility of the adjustable worksurface, and the convenient layout of the items on the workstation. Managers appreciated the flexibility of the workstation, which could be used for a variety of purposes. This particular workstation became the plant standard. Finally, the company medical doctors indicated that the use of ergonomics had eliminated most complaints of back and shoulder strain.

There is one interesting aspect about this case study which we have not addressed, namely the allocation of tasks between automation and manual labour. When the printer assembly was first planned, it was envisioned that the manufacturing would be totally automated. However, it was soon obvious that a few manual workstations were

Table 2.5 Ergonomics of handtools, controls, and displays

1. Use of lightweight power tools
2. Reduction of number of controls on robot control panel
3. Schematic lines on panels to identify the required control sequence
4. Use of colour coding and colour contrast
5. Light fixtures inside control panels to simplify maintenance

necessary for tasks that were difficult to automate. These were workstations where intricate assembly work was performed – 'leftover' tasks which were too complicated for robots to perform.

Several years later, after several product modifications, this assembly is almost totally manual. Most of the automated workstations have disappeared, because of product modifications that were introduced in later models. It was difficult to reprogram the automation and time consuming to acquire new process machinery.

The printer had been designed for ease of assembly, implying that a minimum number of parts are used. It turned out that not only did this simplify the automatic assembly but also the manual assembly. In the end, it became so simple to assemble the printer manually that the cost of automation was not justified.

This case study serves as an example of task allocation between people and automation. The automation failed, which must have disappointed many engineers. There are certainly other less intricate jobs that are easier to automate. For example, in the automobile industry, welding and painting are mostly performed by robots. However, assembly work, with its intricate movements, is more troublesome to automate.

Chapter 3

Anthropometry in Workstation Design

The basic philosophy of ergonomics is to design workstations that are comfortable, convenient and productive to work at. Ideally, workstations should be designed to fit both the body and the mind of the worker. In this chapter we limit ourselves to the body, which certainly is the easier of the two problems. We will demonstrate how adjustability of chairs, stools, benches, and so forth, can help to accommodate people of different body size. By the use of anthropometric design principles it is possible for a variety of people to find physical comfort at a workstation. On the other hand, by not taking into consideration these physical requirements, one may create bad work postures, which lead to fatigue, loss of productivity and sometimes injury.

Anthropometry is not only a concern about appropriate working height, but also about how the operator can easily access controls and input devices. In an automobile it should be possible for a small driver to reach the controls on the dashboard while being held back by the seatbelt. Similarly, the controls of machine tools must be easy to reach. The lathe shown in Figure 3.1 was originally described by Singleton (1962). It is a classic design and makes a clear argument. To control this particular piece of equipment the ideal operator should be 137 cm (4.5 feet) tall, 62 cm (2 feet) across the shoulders, and have a 235 cm (8 feet) armspan, which is closer to the shape of a gorilla!

3.1 Measuring Human Dimensions

There are large differences in body size due to gender and genetics. Men are, on average, 13 cm (5 in.) taller than women and are larger in most other body measures as well.

Genetic differences are evident from a comparison of individuals living in different countries. For example, the average male stature in the USA is 167 cm (66 in.), whereas that in Vietnam is 152 cm (60 in.). A car designed for the US population would fit only about 10% of Vietnamese, unless of course the differences can be compensated for by using an adjustable seat (Chapanis, 1974). However, some of the differences between countries are decreasing, suggesting that there are factors beyond genetics. For example, during the last 20 years the average Japanese teenager has become 2 cm taller (Pheasant, 1986). This is largely attributed to changes in eating habits; in particular, animal proteins have become much more common in the Japanese diet.

According to a study done in the UK, the average male manager is 3–4 cm taller than the male blue-collar worker (Pheasant, 1986). There could be many reasons for this. It may be that taller people are more often promoted to managers, or that taller people are a little more intelligent, or that managers come from a higher social class and thus had better education and also eat more animal protein. It is difficult

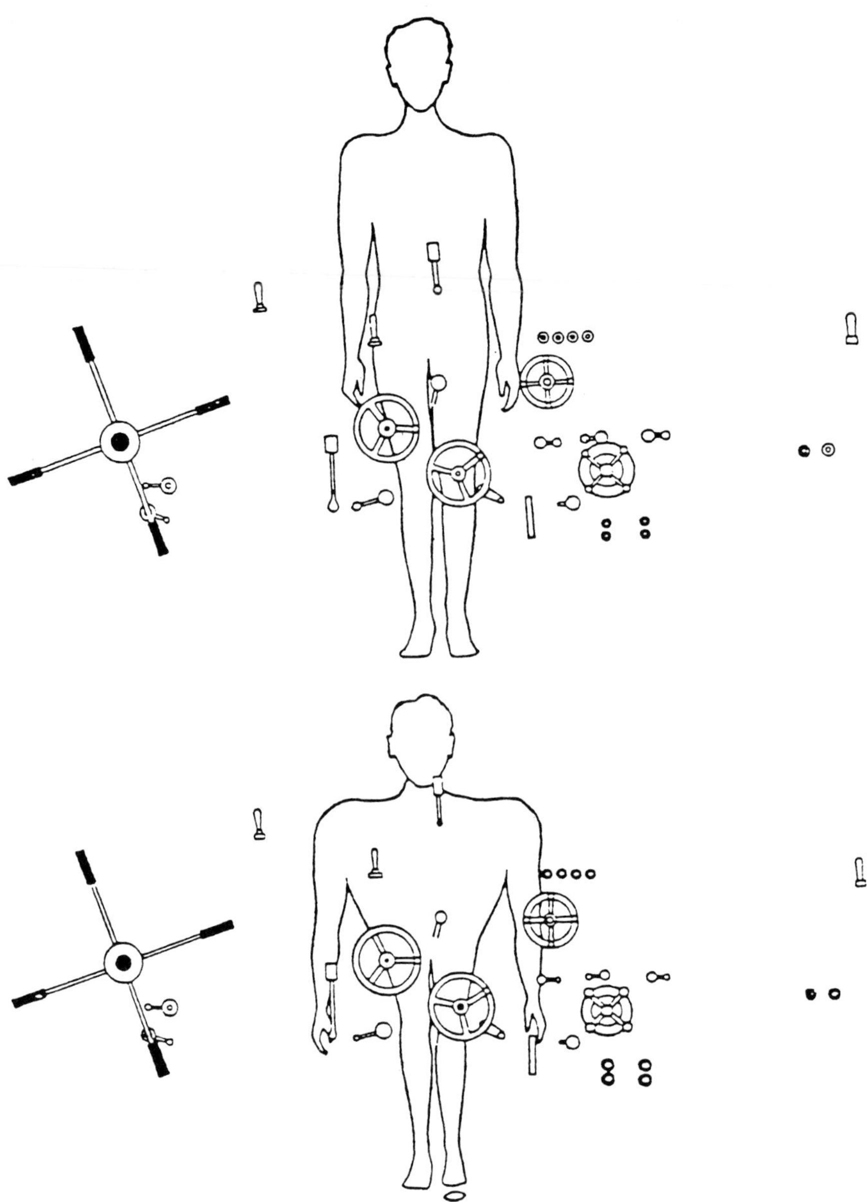

Figure 3.1 The controls of a lathe are not within easy reach of the average man. The bottom figure shows the ideal size operator (Singleton, 1962)

to attribute causes, but probably all of these reasons contribute. Of particular interest to ergonomics is that a male manager may have a different physical frame of reference than the individuals who work for him. For example, a managerial chair is oversized and uncomfortable for a female secretary, and vice versa. A manager may have difficulties in understanding problems related to physical accommodation, simply because they do not apply to him.

Anthropometric measures are usually expressed as percentiles. The most common are the 5th, 50th and 95th percentile measures (Table 3.1). Anthropometric data are usually normally distributed (Figure 3.2) (Roebuck *et al.*, 1975). A normal distribution is characterized by its mean value and its standard deviation (SD). As long as we know these two values of distribution, it is possible to calculate any percentile value. For example, the 95th percentile equals the mean value plus

Table 3.1 Explanation of percentile measures

Percentile	Description
5th	5% of the propulation is smaller
50th	Average value
95th	95% of the population is smaller

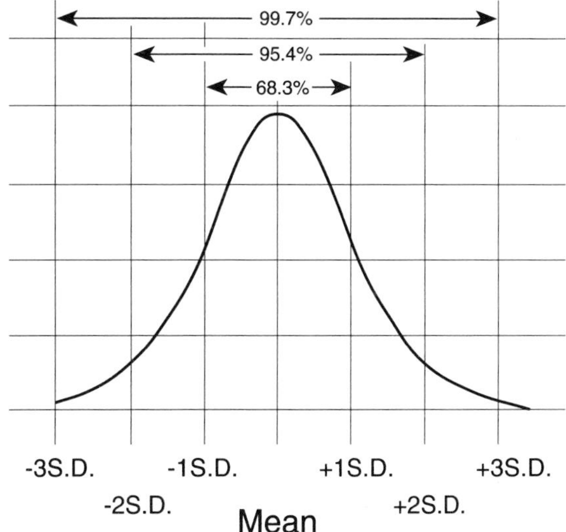

Figure 3.2 Anthropometric data are usually normally distributed

1.65 SD and the 5th percentile equals the mean minus 1.65 SD (Figure 3.2).

The common procedure is to design for a range of population from the 5th percentile (small operator) to the 95th percentile (large operator). The choice of 5th and 95th percentiles is traditional, although one can argue that a greater percentile range should be used. But many ergonomists consider that it is impossible to include extremes of the population, such as dwarfs and giants, in the common design range. For example, the seat of a height-adjustable chair in the USA must adjust between 16 and 20.5 in., which roughly corresponds to the range established by 5th percentile females and 95th percentile males, although some users may be smaller or larger (see Human Factors Society, 1988). Similarly, it would not be practical to make door openings 8 feet tall, although this may be required by a giant.

The greater the design range, the greater the cost. It is more expensive to design for the 5th to 95th percentile range than for the 10th to 90th percentile range. The percentile value selected is largely a political decision, and companies may adopt different policies. One potentially controversial question is whether one should design for the worker population at hand, e.g. the 5th to 95th percentile male, or if one should extend the range to 5th percentile female workers in order to provide 'equal physical access' to females.

There may be reasons to think that workers in a specific manufacturing plant have different body size and are not typical of the population at large. These were the concerns in a study we performed for IBM Corporation in San Jose (Helander and Palanivel, 1990) At this location there were about 1000 female microscope operators, many of whom had recently arrived to the USA from Asia and were shorter than the 5th percentile US female. Many operators had to stretch to be able to get to the eyepiece and they could not put their feet on the floor. We measured 17 different anthropometric measures of 500 operators and calculated the means, standard deviations, and 5th and 95th percentile measures. These measures were used to specify the appropriate measures for the microscope workstation.

We will explain below how anthropometric measures can be translated into workstation design measures by using the 'anthropometrics design motto':

Anthropometrics design motto

- Let the small person reach.
- Let the large person fit.

These principles imply that reach distances should be designed for the small, 5th percentile individual, whereas clearance dimensions should be designed for the large, 95th percentile individual. A simple case of anthropometric design is illustrated in Figure 3.3. The 5th percentile female and 95th percentile male measures are illustrated for a sitting workplace. Note that the popliteal height (from the sole of the foot to the crease under the knee) is 36 cm (14.0 in.) for 5th percentile females and 49 cm (19.2 in.) for 95th percentile males. These values may actually differ slightly in different anthropometric tables. Note also that the popliteal height (and other measures) are taken without shoes, so that for design purposes one must add the height of the heel of the shoe (about 3 cm). The appropriate range of adjustability for a chair-seat height is then 39–52 cm (15–20.2 in.). The distance from the floor to the elbow is obtained by adding the popliteal height, sitting elbow height and shoe height (3 cm). This measure is 57–81 cm (22–32 in.) and it can be used to select appropriate table height.

As illustrated in the right-hand part of Figure 3.3, there are two different ways to compensate for anthropometric differences. One can use either a height-adjustable chair plus a foot rest, or a height-adjustable chair plus a height-adjustable table. Both arrangements will make it possible to support the feet and have the table at elbow height. The height-adjustable table is more expensive than the foot

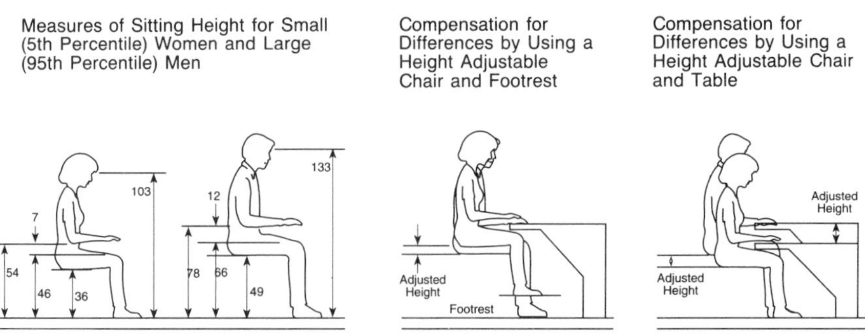

Figure 3.3 Comparison of anthropometric measures (cm) for a sitting 5th percentile female and a sitting 95th percentile male – height-adjustable chairs and tables can be used to compensate for these differences

rest, but it is more comfortable to rest the feet on the floor than to use a foot rest.

In many offices (including the author's) the table height has been set once and for all. The table can indeed be raised and lowered, although not easily. The height-adjustable chair is rarely changed more than once per day. The implications are that for individuals who have their own workstation, ease of adjustability is not crucial. But for people who share a workstation, for example shift workers, adjustability becomes essential (Shute and Starr, 1984). Most of the microscope workers at IBM worked three shifts, so adjustability of the workstation was important. Microscope work is an exacting task. It is necessary to adjust the eye pieces to the exact level of the eyes, the table so that it is convenient to reach microscope controls, and the chair to be able to put the feet on the floor. This is a complex case of adjustability, since there are three interacting elements of adjustability.

3.2 Definition of Anthropometric Measures

The most complete source of anthropometric measures has been published by the US National Aeronautics and Space Administration (NASA, 1978). This reference publication contains measures of 306 different body dimensions, from 91 different populations around the world. About half of the populations are aeroplane pilots, which illustrates the great importance attributed to the anthropometric design of cockpits. Anthropometric investigations have been supported by the Air Force in the USA and many other countries, but surprisingly there is a lack of civilian anthropometric measures. In the USA there has actually never been a comprehensive civilian anthropometric investigation. The measures listed in Table 3.2 are adapted from the data reported by McConville *et al.* (1981), who extrapolated civilian body measures by using data from the military. The measures are also illustrated in Figure 3.4.

Some of the anthropometric measures have Latin names. This is practical, since they refer to bone protrusions on the human body. For example, the tibial height is the height of the proximal medial margin of the tibia, a bone protrusion on the tibia under the knee cap. The acromion height refers to the highest point of the shoulder blade, and the popliteal height is the height from the sole of the foot to the crease under the knee between the upper and the lower leg.

The anthropometric measures illustrate that there are large differences between the sexes. For many measures, the 5th percentile (small male) is about the same size as the 50th percentile (average) female. For example, the inside diameter of the hand grip (measure 17) is 4.3 cm for a 50th percentile female and 4.2 cm for a 5th percentile male. This measure is important for design of handtools to fit the size of the tool to the size of the hand. Women often complain that they have to use handtools designed for men, resulting in muscle fatigue of the hand and the arm, lower productivity, and possibly also injuries (Greenburg and Chaffin, 1977). Both the US Department of Defense as well as industry (e.g. General Motors) have taken note and now supply handtools of different sizes for males and females.

All the measures listed in Table 3.2 have implications for manufacturing. The measurements and their implications are explained below.

1. *Tibial height*. This measure is important for manual materials handling. Items located between the tibial height and the

Table 3.2 US civilian body dimensions (in cm with bare feet; add 3 cm to correct for shoes) of industrial relevance. Adapted from McConville et al. *(1981)*

	Female			Male		
	5th	50th	95th	5th	50th	95th
Standing						
1. Tibial height	38.1	42.0	46.0	41.0	45.6	50.2
2. Knuckle height	64.3	70.2	75.9	69.8	75.4	80.4
3. Elbow height	93.6	101.9	108.8	100.0	109.9	119.0
4. Shoulder (acromion) height	121.1	131.1	141.9	132.3	142.8	152.4
5. Stature	149.5	160.5	171.3	161.8	173.6	184.4
6. Functional overhead reach	185.0	199.2	213.4	195.6	209.6	223.6
Sitting						
7. Functional forward reach	64.0	71.0	79.0	76.3	82.5	88.3
8. Buttock-knee depth	51.8	56.9	62.5	54.0	59.4	64.2
9. Buttock-popliteal depth	43.0	48.1	53.5	44.2	49.5	54.8
10. Popliteal height	35.5	39.8	44.3	39.2	44.2	48.8
11. Thigh clearance	10.6	13.7	17.5	11.4	14.4	17.7
12. Sitting elbow height	18.1	23.3	28.1	19.0	24.3	29.4
13. Sitting eye height	67.5	73.7	78.5	72.6	78.6	84.4
14. Sitting height	78.2	85.0	90.7	84.2	90.6	96.7
15. Hip breadth	31.2	36.4	43.7	30.8	35.4	40.6
16. Elbow-to-elbow breadth	31.5	38.4	49.1	35.0	41.7	50.6
Other dimensions						
17. Grip breadth, inside diameter	4.0	4.3	4.6	4.2	4.8	5.2
18. Interpupillary distance	5.1	5.8	6.5	5.5	6.2	6.8

1 in. = 2.54 cm.

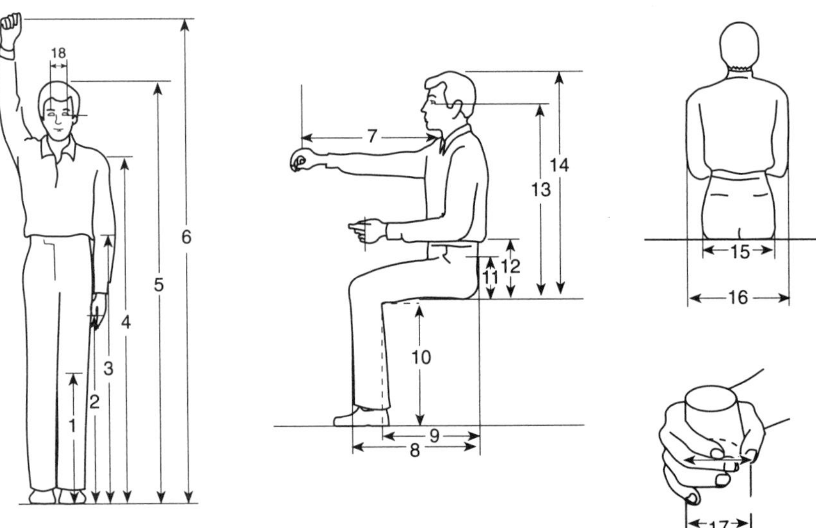

Figure 3.4 Illustration of the anthropometric measures given in Table 3.2

knuckle height must usually be picked up from a stooped position.

2. *Knuckle height*. This height represents the lowest level at which an operator can handle an object without having to bend the knees or the back. The range between the knuckle height and

the shoulder height is ideal for manual materials handling and should be used in industry.

3. *Elbow height*. This is an important marker for determining work height and table height.
4. *Shoulder (acromion) height*. Objects located above shoulder height are difficult to lift, since relatively weaker muscles are employed. There is also an increased risk of dropping items.
5. *Stature*. This is used to determine the minimum overhead clearance required to avoid head collision.
6. *Functional overhead reach*. This is used to determine the maximum height of overhead controls.
7. *Functional forward reach*. Items that are often used within the workstation should be located within the functional reach.
8. *Buttock–knee depth*. This defines the seat depth for chairs and clearance under the work table.
9. *Buttock–popliteal depth*. This is used to determine the length of the seat pad.
10. *Popliteal height*. This is used to determine the range of adjustability for adjustable chairs.
11. *Thigh clearance*. Sitting elbow height and thigh clearance help to define how thick the table top and the top drawer can be.
12. *Sitting elbow height*. Sitting elbow height and popliteal height help to define table height.
13. *Sitting eye height*. Visual displays should be located below the horizontal plane defined by the eye height.
14. *Sitting height*. This is used to determine the vertical clearance required for a seated work posture.
15. *Hip breadth*. This is used to determine the breadth of chairs and whole body access for clearance.
16. *Elbow-to-elbow breadth*. This is used to determine the width of seat backs and the distance between arm rests.
17. *Grip breadth, inside diameter*. This is used to determine the circumference of handtools and the separation of handles.
18. *Interpupillary distance*. This is an important measure in determining the adjustability of eyepieces on microscopes.

To reduce measurement error, anthropometric measures are gathered for minimally clothed men and women who are standing or sitting erect. People in industry are, however, usually fully clothed and stand or sit with a more relaxed posture. With shoes on, the height measures in Table 3.2 should be increased by approximately 3.0 cm. To compensate for postural slump, 2.0 cm is subtracted from standing height and 4.5 cm for sitting height (Brown and Schaum, 1980).

The measure of functional forward reach assumes that there is no bending from the waist or the hips. By bending from the waist, the forward function reach can be increased by about 20 cm and bending from the hips increases reach by about 36 cm (Eastman Kodak, 1983). Since a person cannot bend at the waist or hips for an extended time, these extra allowances should be used only for occasional, short-duration tasks.

There are many different anthropometric databases in use, some of which are fairly dated and may not reflect the fact that the population keeps getting taller. But it may also be the case that some anthropometric data are inaccurate and researchers have not used enough precautions in obtaining accurate measurements. Anthropometric measures are well defined, and there are standard procedures for taking them. There are also special tools and equipment available for taking the measures (see Roebuck *et al.*, 1975).

3.3 Using Anthropometric Measures for Industrial Design

In the past, most research and anthropometric surveys have been initiated by the US Air Force, which presently is developing tools for three-dimensional modelling using computer-aided design. There are already several programs available for computer-aided design including CAR (Crew station Assessment of Reach), SAMMIE (System for Aiding Man–Machine Interaction Evaluation), COMBIMAN (Computerized Biomechanical Man-Model), CREWCHIEF, and ADAM and EVE. A review of these models can be found in Kroemer *et al.*, (1988).

Most anthropometric models have been used to model workstations which involve very tight constraints, such as cockpits. In the industrial environment, there are fewer constraints. People can usually move around freely, and there is not a great need for very sophisticated modelling. Anthropometric design of a workstation can be accomplished in a couple of hours with paper and a pen.

Depending upon the application, anthropometry is used differently (Figure 3.5). In designing cars, it has been common to start with the hip joint or hip reference point (HRP) and then 'laying out' the rest

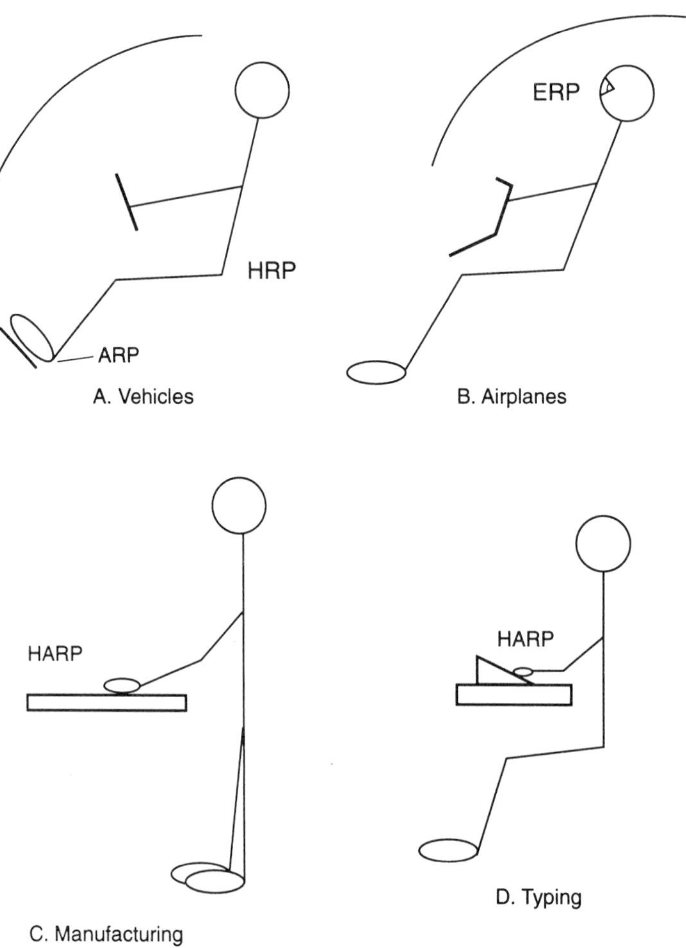

Figure 3.5 Anthropometric design can use different reference points

of the body going up to the head and hands and down to the feet. Some automobile manufacturers (of racing cars) start off with the accelerator reference point (ARP) and then lay out the rest of the body going upwards. In the design of fighter planes, it is important to put the eye at the right height, since there are many displays which must be visible including head-up displays (HUD) which are projected in the windshield. Since the pilot is tied back to the seat, one can make a very accurate estimation of where his or her eye will be. In this case the design will start with the eye or eye reference point (ERP), and the rest of the body can be modelled going downwards.

In manufacturing, for assembly work we advocate the use of a hand reference point (HARP). The ideal location of the hands depends on the task. For heavy manual jobs, the hands should preferably be about 20 cm below elbow height, but for precision tasks with supported under-arms the hands should be about 5 cm above elbow height. Therefore, to design a workstation one first needs to determine the most convenient hand height for the task in question. The rest of the body can then be laid out by finding measures down to the feet and up to the head. Typists have a similar work situation. It may be preferable to start off with a HARP and then lay out the rest of the body.

It is common to design for the range from the 5th to the 95th percentile. In doing so, one may have to add different anthropometric measures. For example, for a sitting workstation with the table top at the elbow height it is necessary to add two measures: popliteal height and sitting elbow height. The addition of anthropometric measures actually produces an inaccurate estimate, since very few individuals are 5th percentile throughout. Typically, a person with a short back may have long legs, or vice versa. Kroemer (1989) showed that the correlation coefficient between stature and sitting eye height is $r=0.73$, between stature and popliteal height $r=0.82$, and between stature and hip breadth is $r=0.37$ ($r=1.00$ is a perfect positive correlation between two measures, $r=0$ implies no relationship between two measures). If two 5th percentile measures are added, the resulting measure could be about the 3rd percentile. And if two 95th percentile measures are added, the resulting measure might be the 97th percentile. This problem (with the table top height) would be solved if there were a single measure for sitting elbow height in the anthropometric tables. However, this measure is not one of the 306 defined in NASA's anthropometric tables (NASA, 1978).

Despite several sources of error in anthropometric data, it is usually possible to estimate anthropometric measures with an accuracy of about 1 cm. This is satisfactory for industry. In fact, individuals sitting at a workstation do not have the sensitivity to judge changes smaller than 1 cm (Helander and Little, 1993). If the chair height is raised or lowered by 1 cm, the chair user will not notice the difference.

3.4 Procedure for Anthropometric Design

A procedure for anthropometric design is presented below.

1. *Characterize the user population*. What anthropometric data are available? Can existing anthropometric data be used with the present population? If there are no valid data, consider creating a database by obtaining measures of the existing workforce.
2. *Determine the percentile range to be accommodated in the workstation design*. If the workforce is dominated by either men

or women it would make sense to design for the predominant sex, for example by using 5th–95th percentile male or 5th–95th percentile female measures. On the other hand, it may be an issue of equality to provide 'accessibility' for the other sex. If so, one would design from the 5th percentile female to the 95th percentile male population.

3. *Let the small person reach and let the large person fit*. Determine reach dimensions (5th percentile) and clearance dimensions (95th percentile) for the work situation that is analysed. An example is given in Figure 3.6. In this manufacturing task, the operator is sitting on a chair with his or her hands at elbow level and manipulating objects 6 cm above the table height. Two important reach measures are the popliteal height from the chair seat to the floor and the buttock–popliteal depth (see Table 3.2). Operators should not sit with dangling feet but should be able to reach the floor. An adjustable chair must therefore adjust to a low level corresponding to the 5th percentile. The buttock–popliteal depth should be the 5th percentile, because if it is longer a small operator will not be able to reach to the back support with his or her buttocks. A clearance dimension (*D*) is created under the table. Assuming the table is height adjustable and can be lowered 10 cm below elbow height, will there then still be enough space for the thighs?
4. *Find the anthropometric measures that correspond to the workstation measures*. The calculations for the 5th percentile female and the 95th percentile male operator are shown in Figure 3.6. The anthropometric measures are added starting from floor level. By using the popliteal height and adding 4 cm for shoes, the required range of seat-height adjustability is calculated to be 39.5–52.5 cm. The sitting elbow height for the 5th percentile

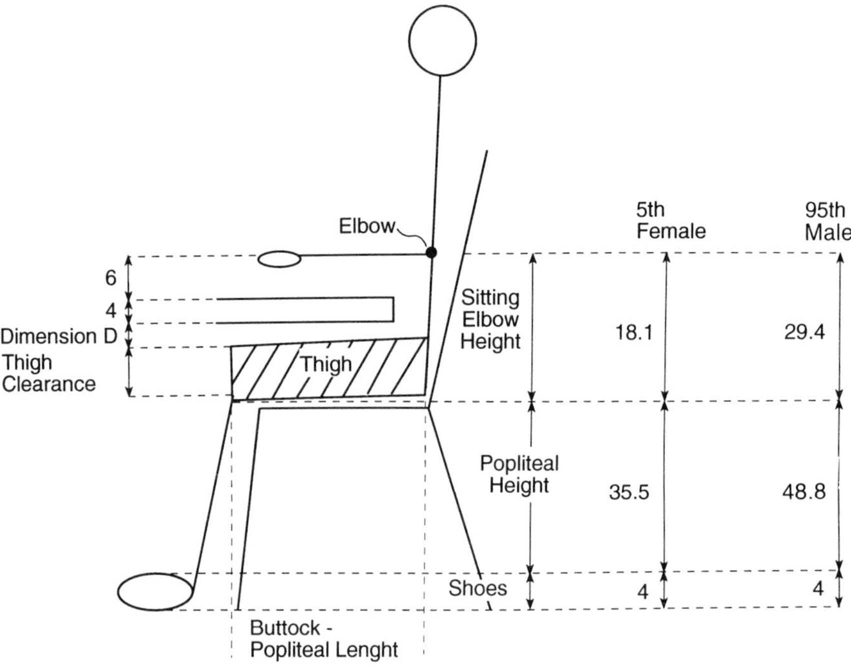

Figure 3.6 Anthropometric measures used to calculate the adjustability of seat height and table height

operator is 18.1 cm and for the 95th percentile operator 29.4 cm. From the sitting elbow height, deduct the thickness of the product (6 cm). This means that the distance from the chair seat to the top of the table is 12.1 cm for the 5th percentile and 23.4 cm for the 95th percentile. Adding these measures to the seat-height adjustability gives a required table height adjustability of 51.6–75.9 cm (or 52–76 cm).

Deducting further, bearing in mind the thickness of the table top, we find that for the 5th percentile there is 8.1 cm of clearance between the chair seat and the table and for the 95th percentile there is 19.4 cm of clearance. Since the thigh clearance (see Table 3.2) is 10.6 and 17.7 cm, respectively, a small female operator will not have enough space, but a large male will be able to fit his legs under the table.

5. It is sometimes difficult to illustrate a work situation using an anthropometric model. Anthropometric measures are static, and in the real world there are many dynamic elements. Operators reach for tools and parts and swing around in the chair. To evaluate the dynamic aspects of a workstation appropriately, one may construct a full-scale mock-up out of cardboard or styrofoam. This will not usually take more than a couple of hours. The purpose is then to have people of different sizes testing out the workstation by moving their body and simulating the task. Through the full-scale mock-up it may be possible to identify features of the workstation which need to be redesigned.

3.4.1 Exercise: Designing a Microscope Workstation

Using the set-up of the microscope workstation shown in Figure 3.7, calculate adjustability ranges for seat height, table-top height, and microscope eyepiece eye height (measures *A*, *B* and *C* in Figure 3.7).

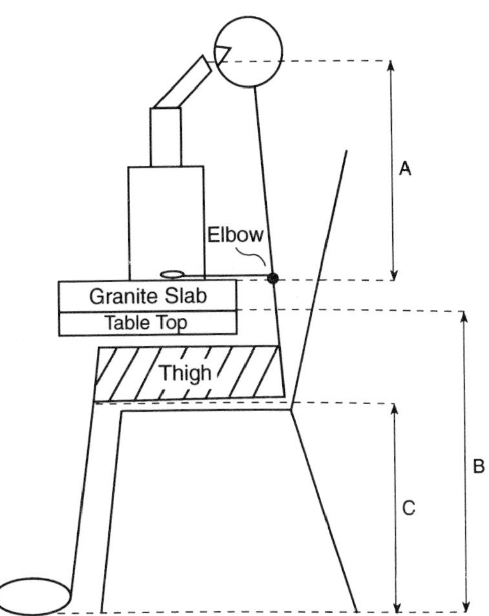

Figure 3.7 Example: designing a microscope workstation

Design for a 5th to 95th percentile female population. There are several assumptions:

1. There is no footrest.
2. The shoes are 4 cm high.
3. In the upper part of the body from the elbow height to the shoulder height there is a postural slump of 2 cm.
4. When looking into the microscope the operators bend the head forward about 30°, which moves the position of the eye downwards by 1.5 cm.
5. The hands are manipulating focusing controls at elbow height.
6. The arms are horizontal and resting on the granite slab.
7. The table top is 3 in. high. There is also a 4 cm thick granite slab on top of the table to reduce vibration.

Answer

Use dimensions listed in Table 3.2.
Measure *A*: 64–75 cm
Measure *C*: 39.4–48.3 cm
Measure *B*: 53.5–72.4 cm.

Chapter 4

Physical Work and Heat Stress

4.1 Physical Workload and Energy Expenditure

In most Western countries physical workload is no longer as common as it used to be. In manufacturing, hard physical labour has been taken over by materials handling aids, mechanized processes and automation. Legislation has also put limits on the workload that employees can be exposed to. Yet, in some occupations such as construction work, commercial fish netting and logging, workers still perform much physical work. Such work generally involves less structured tasks, and they are difficult to mechanize.

Although physical work activities have become less important in Western countries, they are still common in industrially developing countries, where mechanization does not yet pay off. For example, in the construction industry materials are typically carried by workers. Eriksson (1976) estimated that 200 workers at a road construction site in Bangladesh could move as much dirt manually as one Caterpillar, and the costs were equivalent. Under such circumstances, the national economy as well as the workers' private economy will gain by using manual labour. Although the physical work demands in manufacturing have been reduced, there are still situations which require ergonomic analysis. Many individuals are less capable of physical work, and we are particularly interested in individual differences due to gender and age.

4.1.1 Metabolism

Metabolism may be defined as the conversion of foodstuffs into mechanical work and heat (Åstrand and Rodahl, 1986). In order to be useful to the body, the foodstuff is converted into a high-energy compound adenosine triphosphate (ATP). ATP serves as a fuel transport mechanism. It can release chemical energy to fuel internal work in the various body organs. The ATP conversion process is only about 50% efficient, so that half of the total food energy is lost as heat before it can be used. The ATP energy is used in three different processes. First it maintains chemical processes, such as the synthesis and maintenance of high energy bonds in chemical compounds. Second, it is used to fuel neural processes and muscular contractions to maintain the body functions, such as blood flow and breathing. Finally, some of the ATP energy is used for muscular work. At most, 25% of the energy that enters the body in the form of food can be used for muscular work. This is the upper limit of the energy efficiency for the human body, and it is typically achieved for the large muscles in the body, such as the leg muscles. The 25% efficiency exceeds that of a steam engine and is about equal to the efficiency of a combustion engine (Brown and Brengelmann, 1965). For the smaller muscles in the arms and shoulders an efficiency of about 10–15% is fairly typical.

The amount of energy expenditure associated with a task can be assessed by measuring the amount of oxygen used. The *oxygen uptake* is calculated by measuring the volume and oxygen content of exhaled and inhaled air. This analysis is performed using special instruments. The oxygen uptake is then converted into kilocalories of energy expenditure; one litre of oxygen generates 4.83 kcal of energy. Measurement of oxygen uptake therefore provides an exact assessment of energy expenditure, but it is an elaborate procedure. A much easier, but approximate, method is to measure heart rate. Heart rate gives a fair estimate of energy expenditure in the intermediate range. Heart rate is less suitable for assessing small and very high rates of physical work.

Maintaining the basic body function at rest requires about 1200 kcal/day. This is referred to as the basic metabolic rate (BMR). It includes functions such as the heart (215 kcal/day), brain (360 kcal/day), kidney (210 kcal/day), and muscles at rest (360 kcal/day). On top of maintaining the basic body functions, people usually engage in some minimal activity. This is referred to as leisure activity and does not include work activities. Together the BMR and leisure activities give an average energy consumption of 2500 kcal/day.

Different occupations incur different energy consumption rates. For an 8-hour work day the following values are typical:

- Seated office work, 800 kcal/day.
- Light assembly work, 1680 kcal/day.
- Ocean fish netting, 4800 kcal/day.
- Lumberjacking, 6000 kcal/day.

Ocean fish netting and lumberjacking are unusual because of their very high energy requirements.

Total energy requirements are obtained by adding together the BMR, leisure activities and the occupational rates. A total energy requirement of less than 4000 kcal/day is considered moderate, between 4000 and 4500 kcal/day as heavy, and above 4500 kcal/day as severe.

4.1.2 Individual Differences

As noted above, one of the main reasons for taking an interest in work physiology is to consider variations in work capacity between individuals. One important difference is physical condition (Figure 4.1). A highly trained individual (such as a marathon runner) can sustain 50% of the maximal aerobic capacity for an 8-hour work day, an average individual can sustain 35%, and an untrained individual 25% (Michael *et al.*, 1961).

Chronological age is a fairly poor determinant of work capacity. One sure conclusion is that the variability between individuals increases with age. Figure 4.2 shows the maximal oxygen uptake for two individuals from the age of 35 years onwards (Miller and Horvath, 1981). The two curves represent two professors of work physiology (who else would have their maximal oxygen uptake tested so frequently?). From the figure we observe that by the age of 65 years individual A was as fit as ever, whereas individual B had a maximal oxygen uptake of 65% of his high value at the age of 35 years.

4.1.3 Metabolism During Work

Once work has begun, it takes some time for the metabolism to 'catch up' with the energy expenditure of the muscles that are engaged in

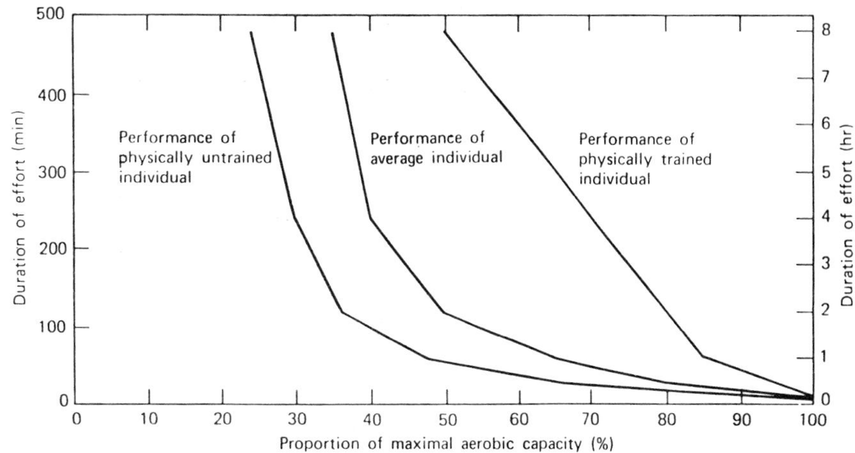

Figure 4.1 The capacity for sustained physical work depends upon the amount of physical conditioning

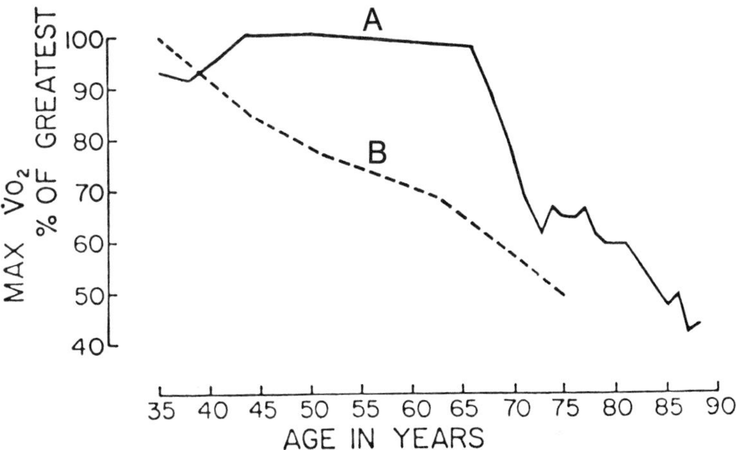

Figure 4.2 Volume of maximal oxygen uptake $\dot{V}O_2$ as a function of age for two individuals. The oxygen uptake is given as a percentage of the greatest value attained for that individual

work. In fact, metabolism does not reach a stable level until several minutes after work has begun. The amount of time taken depends upon how hard the work is, but is typically about 5 minutes. Thus, the metabolic activity (or oxygen uptake) does not increase suddenly at the onset of work. Rather, there is a gradual, smooth increase in oxygen uptake (Figure 4.3). During the initial portion of work, the muscles use a type of energy that does not require oxygen. This type of energy production is known as ‘anaerobic’ (without oxygen) metabolism. A brief task, such as a 100-m sprint uses primarily anaerobic energy.

Anaerobic metabolism is inefficient. It uses nearly 20 times more food fuel than does the aerobic process. It also produces a waste product (lactic acid) which may accumulate in the working muscles rather than being carried away by the blood. Eventually, lack of available energy supplies, lack of fuel, and accumulation of lactic acid in the muscles involved lead to fatigue and cessation of work. It is generally believed that the accumulation of lactic acid results in aching muscles. The same phenomenon is also noted for static work. In this situation (such as carrying a suitcase) static contraction of muscles may produce local muscle fatigue and aching muscles. The accumulation of lactic acid is exacerbated by swelling muscles, which may partially cut off the blood circulation so that the lactic acid cannot be removed effectively.

As the oxygen uptake increases, the body can use the aerobic or oxygen-requiring fuel ATP. Returning to Figure 4.3 it can be seen that the metabolic rate eventually stabilizes. This steady-state level represents the body's aerobic response to the demands of increased workload. When the work ceases, the oxygen uptake returns slowly to the resting level prior to work. During this slow return after work the oxygen debt incurred during the onset of work (area A) is repaid (area B).

4.1.3.1 Example: Calculation of Relative Workload

With a general understanding of the internal energy conversion processes, an example of the calculation of human work efficiency can be discussed. A 30-year-old man of average height (173 cm) and average weight (68 kg) is employed in packaging. This task imposes 23 watts (W) of external work. His resting metabolic rate just prior to work is about 93 W. The steady-state energy expenditure for this task is 209 W. (Both values can be calculated by measuring his oxygen consumption.) The increase in oxygen uptake due to the imposed task is: $209 - 93 = 116$ W. The 23 W of external work therefore imposes 116 W of 'internal work', and the energy efficiency is $23/116 = 20\%$.

The VO_2max. (volume of maximal oxygen uptake) for this 30-year-old man is 3.5 l/min. The oxygen uptake can be converted directly to work, and 3.5 l/min corresponds to 1179 W of work. Assuming a 20% efficiency in energy conversion, this translates to 236 W of external work. The assembly work therefore corresponds to a

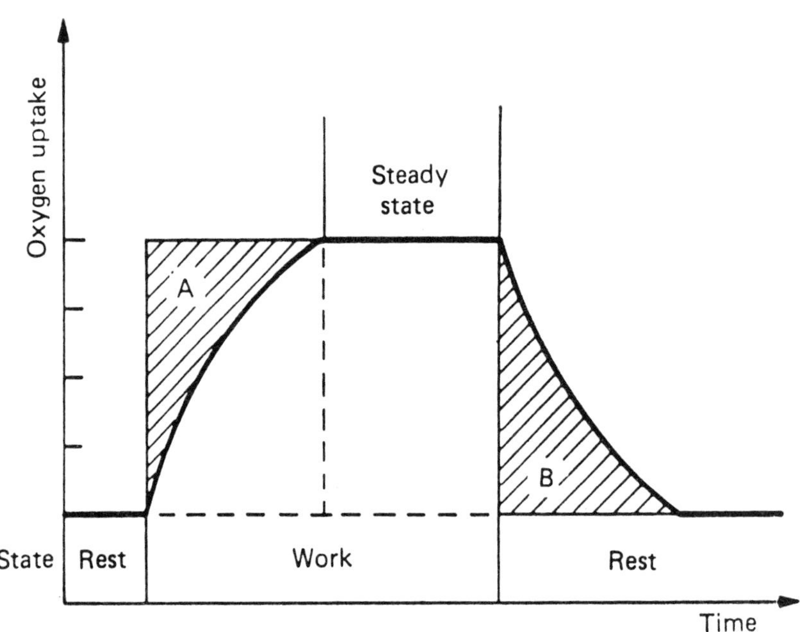

Figure 4.3 Oxygen uptake at the onset of, during and after work. A, oxygen debt; B, repayment of oxygen debt during rest. A = B

23/226 = 9.7% relative load. Compared with Figure 4.1, this is much below 25%, and is not excessive.

This calculation example can be expanded by analysing other individuals with a lower maximal oxygen uptake. For example, a 60-year-old female has a $\dot{V}O_2$max. of 2.2 l/min (Åstrand, 1969). This translates to 134 W of external work and a relative workload of 17%. For an untrained individual with a maximum workload of 25% (see Figure 4.1) this value would be on the high side.

4.1.4 Measurement of Physical Workload

As we have previously noted, it is mostly impractical to use oxygen uptake to assess workload in a manufacturing situation. Heart rate (pulse rate) is a far easier measure. However, heart rate is a good predictor only of workloads of intermediate intensity (about 100–140 beats/min). Simple measurements of heart rate can be useful to estimate if there are any problems with the current level of physical workload. This is illustrated by the following example.

4.1.4.1 Example: Fatigue Due to Physical Workload

The author once visited an automobile assembly plant. There was a female assembly worker who seemed physically exhausted. She was about 45 years of age and of small stature (about 150 cm (5 ft)). The type of work did not seem to put overly great demands on any of her co-workers. However, I stepped up and asked to take her pulse rate. It was running at about 135 beats/min, clearly excessive for an 8-hour work day. She was moved to another, less physically demanding task.

4.2 Heat Stress

Heat stress is often a serious problem in industrially developing countries where work is conducted outdoors or manufacturing facilities lack insulation and/or cooling. Surprisingly, it is often also a problem in southern Europe and the USA. In this section I briefly review the standards on heat stress that have been issued by the International Standards Organization. In no other field of ergonomics are there as many detailed regulations.

4.2.1 Thermoregulation

There are several physiological mechanisms for regulating body temperature. These are under involuntary control by nerve cells in the hypothalamus (a structure in the lower brain), and they maintain the body temperature within a narrow range (about $37 \pm 0.5°C$). This process is known as 'thermoregulation'. As illustrated in Chapter 15, the body temperature exhibits daily variations. It peaks in the late afternoon and reaches its lowest level in the early morning. In order to keep the body temperature within a narrow regulated range, the amount of heat gained and lost by the body over the short span of time must be equivalent. If the body gains an excessive amount of heat, there could be excessive sweating, dehydration, heat stroke and, finally, death may occur.

There are two major ways of adapting to a hot environment: through acclimation and acclimatization (Miller and Horvath, 1981). *Acclimation* refers to physiological changes, such as sweating, in response to temperature. *Acclimatization* refers to more enduring changes in physiological mechanisms that enable an individual to work in extremely hot environments. Repeated exposure to hot environments leads to an improved tolerance to the heat load. During acclimatization there are progressive increases in body temperature, working heart rate and sweat rate. These processes can be completed in 1–10 days of exposure to a hot environment. The time required for acclimatization is reduced when people actually perform physical work in the heat. However, acclimatization to a hot environment can be lost over a period as short as a weekend. People who work outdoors and spend the weekend in an air-conditioned environment will have to acclimatize again. Recovery to the prior level will take about a day. Acclimatization is usually completely lost after 3–4 weeks in a cool environment.

4.2.2 Measurement of Heat Exposure

In addition to the ambient temperature, there are several other factors that effect heat exposure. In order to calculate their effect, the thermal balance of the body may be expressed in the *thermal balance equation*. A somewhat simplified version of this equation is (in $W\ m^{-2}$):

$$M - W = C + R + E + S$$

where M is the metabolic power, W is the effective mechanical power, C is the heat exchange by convection, R is the heat flow by radiation at the skin surface, E is the heat flow by evaporation at the skin surface, and S is the heat storage.

As explained above, the metabolic processes are only partially effective. For the most effective muscles only about 25% of the metabolism (M) can be used for work (W), the rest being used to produce heat and maintain the basic metabolic processes. By expressing the metabolic power in watts per square metre, it is possible to compensate for the body size of individuals. For the calculation of an average individual, one can assume a body area of $1.8\ m^2$.

Heat transfer by convection (C) refers to the temperature exchange produced by moving air. The amount of convection depends on the difference between skin temperature and air temperature. The radiated heat (R) may be heat radiated by the human body (in the infrared light spectrum). The human body can also absorb radiated heat from external sources. The evaporated heat loss (E) occurs primarily at the skin surface. Moisture is present on the skin because of sweating, and when the moisture evaporates heat is taken from the body surface. The evaporation is a function of air speed and the difference in vapour pressure between the sweat (at skin temperature) and the air. In hot, moist environments, evaporated heat loss is limited by the low capacity of the ambient air to accept additional moisture. In 100% humidity there is no evaporation, which limits the cooling of the body (Miller and Horvath, 1981). In a hot, dry environment, however, evaporated heat loss is limited only by the amount of perspiration that can be produced by the worker. The maximum sweat production that can be maintained by an average man throughout a day is $1\ l\ h^{-1}$. The heat storage (S) should in essence balance at around zero. If S becomes large there is a risk of heat stroke. There are obviously many ways to reduce S – stopping working is one way. Several additional methods are mentioned below in Section 4.2.4.

The metabolic rate for different tasks can now be classified as in Table 4.1 (International Standards Organization, 1989a).

4.2.3 Wet Bulb Globe Temperature

One common method of evaluating heat stress is to record the wet bulb globe temperature (WBGT) (International Standards Organization, 1989b). This index takes into account four basic parameters: air

Table 4.1 Classification of industrial activities in terms of workload and metabolic rate

Activity	Workload	Metabolic rate ($W m^{-2}$)
Seated, relaxed	Resting	58
Standing, light industry	Low	93
Standing, machine work	Low	116
Heavy machine work	Moderate	165
Carrying heavy material	High	230

temperature, mean radiant temperature, air speed, and absolute humidity. There are two different formulations for WBGT.

(1) Inside buildings and outside buildings where there is no sunshine:

$$WBGT = 0.7\ T_{NW} + 0.3\ T_G$$

(2) Outside buildings with solar load:

$$WBGT = 0.7\ T_{NW} + 0.2\ T_G + 0.1\ T_A$$

where T_{NW} is the natural wet bulb temperature, T_G is the globe temperature, and T_A is the dry bulb temperature.

These measurements are easy to obtain using the instrumentation illustrated in Figure 4.4. The values of WBGT can now be used to classify the amount of work and to suggest limits for exposure to heat stress (International Standards Organization, 1989) (Table 4.2).

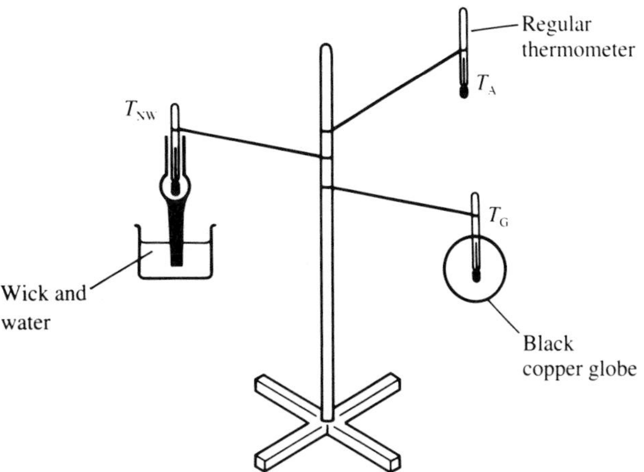

Figure 4.4 The globe temperature (T_G) is measured with a thermometer inside a black painted copper globe; the wet bulb temperature (T_{NW}) is measured with a thermometer put in a wick, the lower part of which is immersed in a reservoir of water; the dry bulb air temperature (T_A) is measured using an ordinary thermometer

4.2.4 Heat Stress Management

A number of ways of reducing heat stress in the work environment are listed below. Note that in each case the practicality of measures must be evaluated.

- Reduce the relative humidity by using dehumidifiers.
- Increase air movement by using fans or air conditioners.

- Remove heavy clothing; permit loose-fitting wide clothing.
- Provide for lower energy expenditure levels.
- Schedule frequent rest pauses; rotate personnel.
- Schedule outside work so as to avoid high-temperature periods.
- Select personnel who can tolerate extreme heat.
- Permit gradual acclimatization to outdoor heat (2 weeks).
- Supply cool, refrigerated vests (containing cooling elements).
- Install local cold spots, e.g. refrigerated rooms for rest breaks.
- Maintain hydration by drinking water and taking salt tablets.

Table 4.2 Reference values of the WBGT heat stress index (adapted from International Standards Organization, 1989)

Workload	Metabolism, *M* ($W\ m^{-2}$)	Reference value of WBGT (°C)	
		Acclimatized	Not acclimatized
Resting	<65	33	32
Low	65–130	30	29
Moderate	130–200	28	26
High	200–260	26	23
Very high	>260	24	19

4.2.5 Comfort Climate

During the past 20 years there has been increasing debate concerning the maintenance of a pleasant climate in office environments. In order to measure the *thermocomfort* under these circumstances, an index called the predictive mean vote (PMV) is used. The PMV is an index that predicts the mean value of the votes that would be obtained if a large group of persons were asked to evaluate the climate. The following seven-point thermal sensation scale is used:

+3 Hot
+2 Warm
+1 Slightly warm
0 Neutral
−1 Slightly cool
−2 Cool
−3 Cold

The PMV can be used to predict the percentage of dissatisfied office users (PPD) (Figure 4.5).

The results of research gives credence to the saying: 'You can't please everybody'. Regardless of the temperature setting in an office there is always at least 5% of the office workers who are going to be dissatisfied. The International Standards Organization suggests that the temperature be chosen so that the PPD is less than 10%, that is 90% of office users like the climate. During the winter season this translates to an indoor temperature of 20–24°C, and during the summer season to an indoor temperature of 23–26°C. Both these temperature ranges assume sedentary activities, such as are common in an office environment. The reason for the lower temperature range during the winter is that thicker, more insulating clothes are worn during winter time. The ISO 7730 points out that there is not sufficient information available to establish comfort limits for activities that are more physically demanding than seated office work (International Standards Organization, 1984).

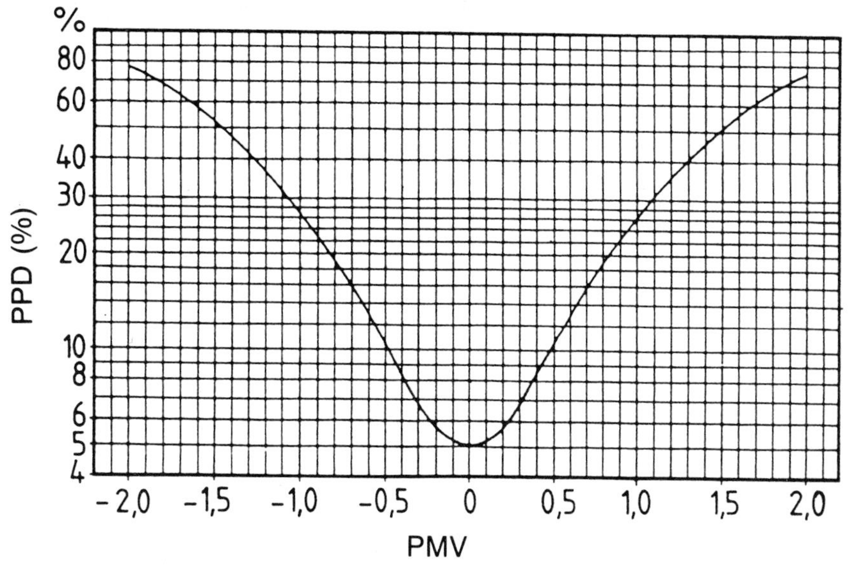

Figure 4.5 The predicted percentage of dissatisfied (PPD) users as a function of the predictive mean vote (PMV) (International Standards Organization, 1984)

4.2.5.1 Example: Discussion of Heat Stress Measures

Discuss the different measures listed under section 4.2.4 that can be taken to reduce heat stress. For each one factor, indicate what entity in the thermal balance equation is being affected.

Chapter 5

Manual Lifting

Low back injuries associated with manual lifting activities are frequent in industry. According to a US Department of Labor Report (1982), back injuries accounted for nearly 20% of all injuries and illnesses in the workplace, and nearly 25% of the annual worker compensation payments. In the USA for the year 1989, Webster and Snook (1994a) estimated the cost of worker compensation for low back pain to be $11.4 billion. Similar figures are obtained from other countries. In the UK, for 1988–1989, 27% of all reported accidents involved manual handling. The annual cost of sickness absence due to back pain and back injuries in the UK is thought to be of the order of £3000 million (Pheasant and Stubbs, 1992).

Manual handling and lifting are the major causes of work-related back pain (Keyserling and Chaffin, 1986). However, back pain, and in particular low back pain, is also common in other work environments such as seated work where there is no lifting or manual handling (Lawrence, 1955). In fact, back pain is extremely common. During a lifetime, there is a 70% chance of developing low back pain, and there is a 1:7 chance that any individual will presently be suffering from back pain (Pheasant and Stubbs, 1992). Many low back injuries seem to happen spontaneously, however, and Magora (1974) indicated that lifting and bending were related to only about one-third of back injuries. Thus, the prevention of back injuries due to lifting, will prevent only a small proportion of injuries.

We will first present statistics of back injuries associated with lifting. We will then analyse 'correct lifting techniques', and what can be done with training to help individuals lift correctly. A biomechanical model for calculating the compressive force in the lower back is presented. This model has been important in establishing federal guidelines for lifting, such as the current directives for the European Community and the NIOSH guidelines in the USA. Finally, several lifting aids that can be used in manufacturing are described.

5.1 Statistics of Back Injuries Associated with Lifting

In 1982, the US Department of Labor published a report of 906 back injuries associated with lifting. The interesting aspect about this study was that only accidents due to manual materials handling were analysed, and there were no 'faking' accidents in the data. From the data it was observed that 42% of all back injuries due to lifting occurred in manufacturing. This was three times as much as for any other industry. Back injuries are therefore frequent in manufacturing and it is important to analyse their causes.

The 906 workers were asked what they were doing when they injured their back (Table 5.1). The percentage values in Table 5.1 add up to more than 100% because many workers reported engaging in more than one activity. In the table we report the number and percentage

Table 5.1 Workers' account for what activities they were engaged in while lifting (some workers reported more than one activity)

Activity	No. of accidents	Activity (%)
Carrying	133	15
Holding	96	11
Lifting	692	77
Lowering	107	12
Placing	145	16
Pulling	65	7
Pushing	39	4
Shovelling	14	2
Other	25	3
Total	906	

of accidents rather than the accident *rate*. The accident rate would be obtained if the number of accidents were divided by the amount of time engaged in each activity. Accident rate would be a much more informative measure. For example, if the individuals who engage in, say, pushing; (4% of the accidents) took only 1% of the total time, then the accident rate for pushing would be 4 times greater than the average. Unfortunately, there is no information on the amount of time that workers spend on each of the different activities. We can therefore only analyse the data from the 'numbers point of view'. From Table 5.1 we conclude that it makes sense to focus on lifting, since a reduction in the number of lifting accidents would have the greatest impact on overall safety.

Workers were then asked what types of movements they were doing when their back was injured (Table 5.2). Again, many workers reported several simultaneous activities. The most dangerous activity was bending, followed by twisting and turning. This verifies what has been pointed out by many researchers – a combination of bending and twisting/turning puts a torque on the spine and the likelihood for back injuries increases.

Workers were asked if lifting equipment was available. The responses are shown in Table 5.3. In the majority of accidents lifting

Table 5.2 Common movements being undertaken when the back was injured

Activity	No. of accidents	%
Bending	505	56
Climbing	16	2
Squatting	107	12
Standing	243	27
Stretching	141	16
Suddenly changing position	159	18
Twisting/turning	299	33
Walking	72	8
Other	26	3
Total	894	

Table 5.3 Workers' response to the question 'Was lifting equipment available'

Response	No. of accidents	%
Equipment not available	434	60
Equipment available but not used because:		
Did not think it was necessary	61	9
It was not practical to use	121	17
It was not working	11	2
It takes too long	16	2
Injury occurred while using equipment	41	6
Other	34	5
Total	708	101

equipment was not available. However, the availability of lifting equipment does not necessarily mean that it will be used. In many cases existing equipment was not used because it was not practical (17%) or workers did not think its use was necessary (9%). The practicality of lifting aids is crucial. If workers find that lifting aids slow down work, the chances are that they will not be used. A slow lifting aid reduces productivity and a worker's sense of accomplishment. It is essential that before ordering lifting equipment, an analysis is made of the practicality of the equipment with regard to the task and the effect on productivity.

5.2 A Biomechanical Model for Lifting

The basic problem in lifting is that the force from a lifted load is multiplied by about 10 times in the spine. We will explain this phenomenon below. The human spine is a flexible column of 24 vertebrae and a large wedge-shaped bone at the bottom, which is called the sacrum (Figure 5.1). Between each pair of vertebrae there are discs which act as shock absorbers. On top of the sacrum are five lumbar vertebrae referred to as L1 to L5. The bottom disc L4/L5 incurs most of the back injuries.

A disc has a fibrous outer layer and is filled with fluid. With increasing age, and also with increasing exposure to manual material handling, cracks develop in the disc, and if there is a great amount of pressure, there is a risk of disc herniation (Michel and Helander, 1994). The fluid of the disc will press through the fibrous outer layer and put pressure on the nerves adjacent to the spine. A graphic but not quite accurate analogy is 'squeezing a jelly doughnut'. Most medical experts now believe that only about 5% of back injuries involve damage to the discs. However, when they occur, these injuries tend to be more serious and long lasting. Fracture of the vertebrae is very rare in lifting accidents, unless the bones have become softened, such as in osteoporosis.

Since disc injuries occur in either the L4/L5 disc or L5/S1 disc, the biomechanical calculations are done for these discs. To calculate the compressive force on disc L5/S1 we make several assumptions. We assume that the individual weighs 75 kg, and that 65% of the body mass is in the upper part of the body, denoted by the vector **B** (Lindh, 1980). The length of the moment arm from the erector spinae muscle to the disc is 6 cm. The calculations are illustrated in Figure 5.2.

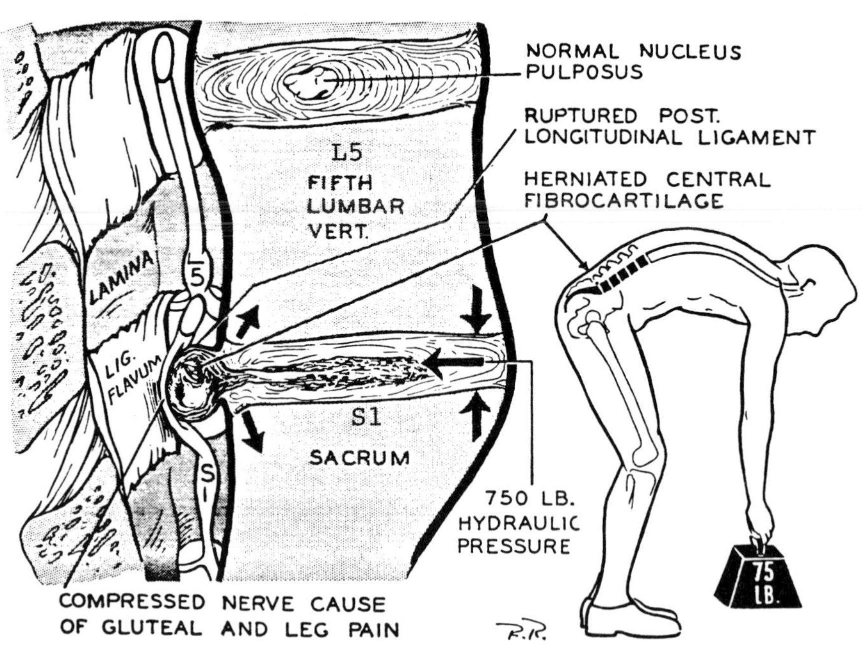

Figure 5.1 Illustration of the L4/L5 and the L5/S1 discs. The L5/S1 disc shows herniation due to lifting. (From Keegan, 1953)

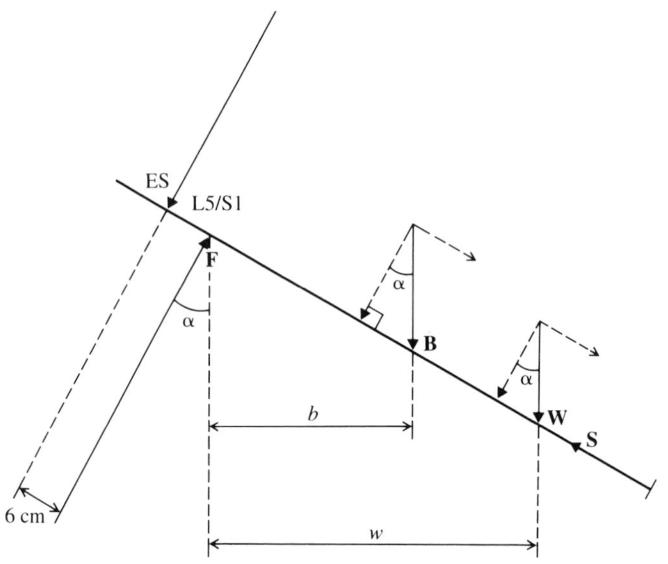

Equilibrium equations:

(1) Moment around L5/S1:

$$\text{ES} \times 0.06 = \mathbf{B}\cdot b + \mathbf{W}\cdot w$$

(2) Add forces in spine direction:

$$\mathbf{F} - \text{ES} - \mathbf{B}\cos\alpha - \mathbf{W}\cos\alpha = 0$$

(3) Add forces perpendicular to spine:

$$\mathbf{S} - \mathbf{B}\sin\alpha - \mathbf{W}\sin\alpha = 0$$

*Figure 5.2 Calculation of the disc compressive force **F** and disc shear force **S**. **ES**, Erector spinae muscle force; **B**, force from upper body weight; **W**, force from lifted weight*

Let us apply this model to the two different cases of lifting shown in Figure 5.3. Assume that for the case of lifting with a bent back (A) the moment arms are $w = 40$ cm and $b = 25$ cm. For lifting with a straight back (B) the moment arms are somewhat reduced: $w = 35$ cm and $b = 18$ cm.

Assuming that **B** = 75x0.65xg = 75x0.65x9.81 = 478 N and that **W** = 250 N, we can use equation (1) to calculate that for case (A) **ES** = 3658 N. Assuming a body inclination of 30°, the disc compressive force is calculated using equation (2): **F** = 3658 + 478x0.89 + 250x0.89 = 4306 N. Similarly, for case (B) **ES** is calculated to 2892 N and, assuming a body inclination of 30°, **F** = 3540 N. This corresponds to a reduction in disc compressive force by 18% for the case with bent knees (B).

This model makes many simplifying assumptions. In the first place lifting is analysed as a static activity, whereas in reality it is very much a dynamic activity. Dynamic lifting models have been developed, and these give compressive disc forces that are 20–200% of the static case (e.g. Garg *et al.*, 1982; McGill and Norman, 1986). These models are still under development, and are not yet practical to use. Chaffin (1969) assumed that abdominal pressure will also effect the lifting model. This additional assumption may, however, not affect the calculations very much (Waters *et al.*, 1993).

5.3 The So-called 'Correct Lifting Technique'

In many organizations training courses are given to instruct employees in correct lifting techniques. This entails lifting with a straight back and bent knees which, as we have seen above, can reduce the disc compressive forces. The International Labour Organization (1972) published several 'kinetic methods' which build upon this technique (Figure 5.4).

The guidelines for correct lifting techniques 'straight back–bent knees' have become quite controversial in the last couple of years. The first observation is that this technique only applies to small compact objects that can be held between the legs while lifting. Larger boxes, for example, are too large to lift with the straight back–bent knees technique. To clear the knees it will be necessary to hold the box at some distance. In addition, there is much more load on the

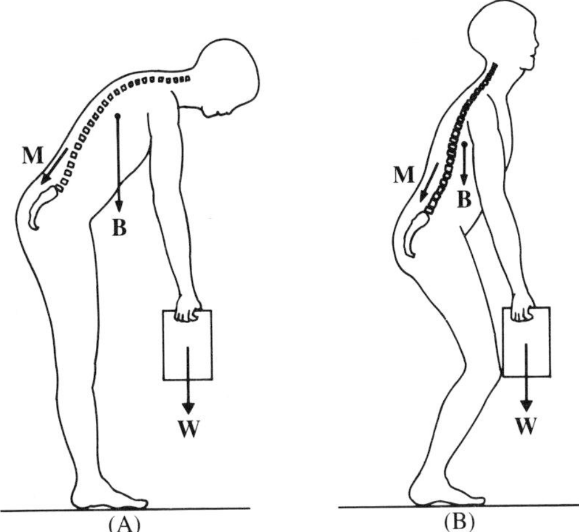

Figure 5.3 An individual weighing 75 kg is lifting an object of 25 kg. (A) Lifting with a bent back. (B) Lifting with bent knees and a straight back

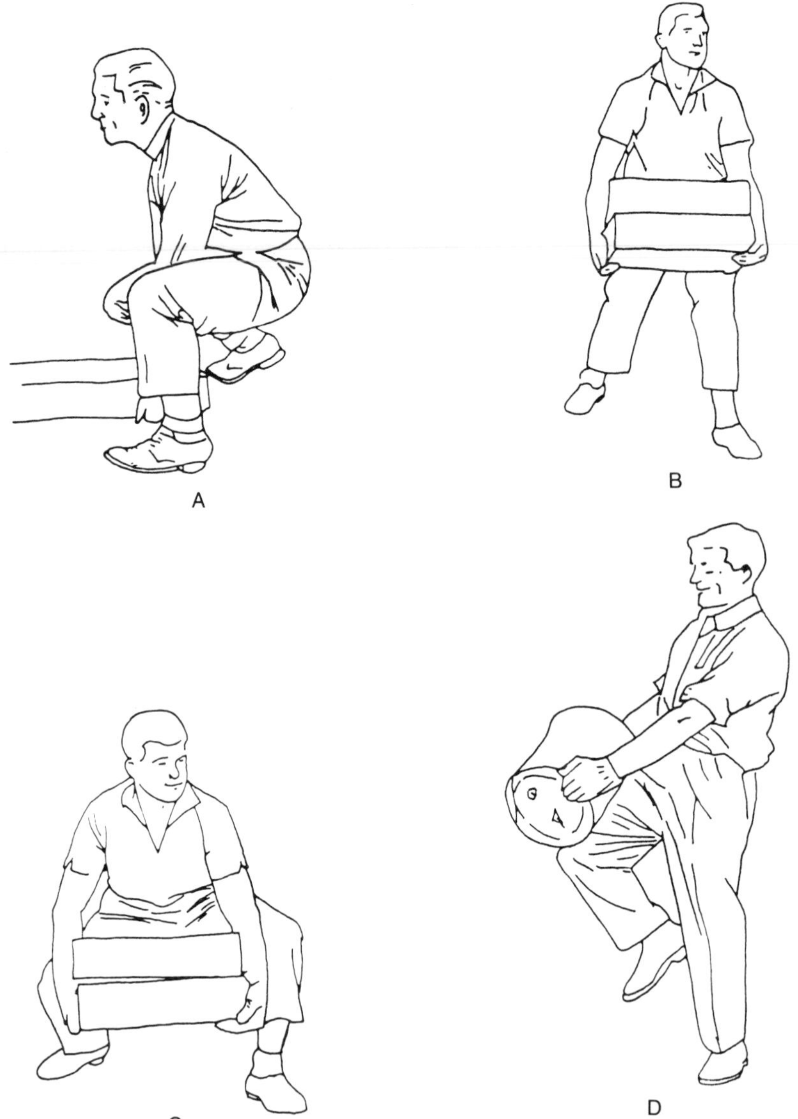

Figure 5.4 Illustration of 'correct lifting techniques'. (A) *Squatting while lifting. The angle of the knee of the front leg is approximately 90°, the arms are held close to the body, and the back is straight. Before raising the load the chin is tucked in, which tends to further straighten the back while lifting. After raising the load the lifter is immediately ready to move horizontally by using the momentum of the body weight.* (B) *A weight should be carried with straight arms. This reduces the tension in the upper arm and shoulder muscles.* (C) *The arms should remain straight while lifting. The feet are placed apart to prepare for forward movement.* (D) *This illustrates how a load can be raised to bench height by using leg muscles, thus reducing the risk of back strain (International Labour Organization, 1972)*

leg muscles. For many lifts this technique is simply difficult and awkward and it will not be used by workers. Furthermore, Garg and Herrin (1979) calculated disc compressive forces and concluded that stoop lifting (bent back–straight knees) is sometimes superior to squat lifting (straight back–bent knees).

There have been many studies investigating the effect of training in manual materials handling. Unfortunately very few studies have used control groups, so it is not possible to draw any firm conclusions. (For a complete discussion of these issues, see Kroemer *et al*. (1994).) The problem with not having a control group is that other simultaneous changes in a company, or among those who are tested, could effect the outcome of training. Let us assume, for example, that training in manual materials handling is introduced in a company. As this is done, not only is awareness of good lifting techniques promoted, but there are also several other simultaneous effects.

- Understanding that low back pain is the cause of elevated workmen's compensation premiums in the company.
- Understanding that the company expects fewer accidents to be reported in the future.
- Understanding the priority expressed by management to reduce low back injuries.
- Feeling better about the personal concern shown by the company compared with in the past.
- Experiencing greater job satisfaction and cohesion with co-workers.

These secondary factors are likely to affect the reporting of accidents. Thus, there are reasons why the reported injury rate could be affected by parameters other than the incidence of low back pain. In a similar fashion, a worker's decision to return to work after low back pain treatment is affected by management attitudes and psychological factors (Snook, 1988).

Assuming one can control for these motivational effects, it could then be possible to make a fair comparison. Scholey and Hair (1989) investigated the incidence of back pain among 212 physical therapists who were involved in back-care education. One would think that physical therapists would be careful to report all occurring back pain and that they would be less biased by secondary factors. Their incidence rate was compared with that in a carefully matched group consisting of individuals who were not physical therapists. There was no difference in the incidence rate reported in the two groups.

In developing a training programme one must first consider what to train (Table 5.4). There are many problems with training programmes (Kroemer *et al*., 1994).

1. There is usually a limited time effect of training. During the immediate time following training, trainees have a sense of enthusiasm and relevance. After a few weeks the information 'sinks back' and is perceived as secondary to many other problems. People tend to revert to previous habits if training is not reinforced.
2. One of the problems in teaching correct lifting techniques is the lack of feedback from the body itself while lifting. There are no nerve endings in the discs, which means that the lifter is not aware of differences in disc pressure due to lifting technique. The trainee

Table 5.4 Options for training programmes in manual materials handling (adapted from Kroemer et al., *1994)*

Lifting skills: correct body positioning, posture and movement
Awareness and attitudes: physics and biomechanics of lifting
Fitness and strength Asian style

must then rely on feedback from the training instructor, peers and managers.

3. Emergency situations which lead to back injury are difficult for an individual to control. As with other accidents, several different things occur simultaneously. The individual must make quick decisions, and body movements cannot be controlled. The situation is quite different from planned, deliberate lifting – which can be controlled. Therefore, if job requirements are basically stressful, behaviour modification through training may not be successful. It is better to design safe jobs, where manual handling is less frequent.

5.4 Guidelines and Standards for Lifting

In many countries there are guidelines and standards which limit lifting in the workplace. The purpose of these guidelines is to reduce the amount of low back pain as well as work injuries. The rationale is that manual lifting poses a risk of low back pain, and low back pain is more likely to occur if the load exceeds the worker's physical capabilities. In addition, the physical capabilities of workers vary extensively, and in designing workplaces and tasks one must consider that some workers are less capable than others.

Below we look at three important sets of guidelines: the NIOSH guidelines (Waters *et al.*, 1993), the European Community guidelines, and the Safety and Health Commission Guidelines, UK. Each takes a different approach to the determination of acceptable weights. Mital *et al.* (1993) provide detailed information on these and other regulations pertinent to manual materials handling.

5.4.1 1991 NIOSH Equation for Evaluation of Manual Lifting

This new equation replaces the former NIOSH lifting equation published in 1981. In developing the present guidelines, three criteria for lifting were considered: biomechanical, physiological and psychophysical (Waters *et al.*, 1993). The *biomechanical criterion* was based on calculating the compressive forces in the L5/S1 disc. Several studies have indicated that, during lifting, the largest moments are created in the trunk area and the L5/S1 disc is at greatest risk. This criterion is most important to delimit the weight of *infrequent, heavy lifts* in the lifting equation. Based on studies of human cadavers it was concluded that a maximum disc compressive force is 3.4 kN, although for some individuals it may be twice as much.

The *physiological criterion* evaluates the metabolic stress and muscle fatigue that may develop during lifting. This criterion is most important for *frequent lifting*. To limit muscle fatigue the maximum aerobic work was set to 9.5 kcal/min. This corresponds to the average, 50th percentile female work capacity. A single lifting task should not impose greater demands than 70% of the maximum aerobic capacity. For long work periods such as 1 hour, 1–2 hours and 2–8 hours, the maximum work rate must be lowered to 50%, 40% and 33% of the maximum aerobic capacity, respectively (see Figure 4.1).

In developing the equation is was considered that working at waist level, at a height of 75 cm, is the most comfortable. Lifts above waist level involve both the shoulder and the arm, whereas lifts below waist level involve the whole body.

The third criterion, the *psychophysical criterion* took into consideration the *acceptability* of lifts to workers. This type of criterion is based on subjective judgment among workers, and the chosen limit for lifting should be acceptable to 75% of female workers and 99%

of male workers. The calculations are based on experimental studies where subjects are asked over the course of an experiment to rate the acceptability of a lifting task.

The NIOSH equation for calculating the recommended weight limit (RWL) represents a compromise between the three different criteria discussed above. It is a multiplicative model and several task variables are included as weighting functions (Waters *et al.*, 1993):

$$RWL = LC \times HM \times VM \times DM \times AM \times FM \times CM$$

Multipliers		*Metric*	*US customary*
LC	Load constant	23 kg	51 lb
HM	Horizontal multiplier	$(25/H)$	$(10/H)$
VM	Vertical multiplier	$(1-0.003\|V-75\|)$	$(1-0.0075\|V-30\|)$
DM	Distance multiplier	$(0.82+4.5/D)$	$(0.82+1.8/D)$
AM	Asymmetric multiplier	$(1-0.0032A)$	$(1-0.0032A)$
FM	Frequency multiplier, obtained from Table 5.5		
CM	Coupling multiplier, varies between 1.00 (good) to 0.90 (poor)		

Variables

- H Horizontal location of hands from the midpoint between the ankles. Measure at the origin and the destination of the lift (cm or in.). H is between 25 cm (10 in.) and 63 cm (25 in.). Most objects cannot be lifted closer than 25 cm from the ankles.
- V Vertical location of the hands from the floor. Measure at both the origin and the end-point of the lift.
- D Vertical travel distance between the origin and the destination of the lift (cm or in.).
- A Angle of asymmetry – angular displacement of the load from the

Table 5.5 Frequency multipliers (cm) 75 cm = 30 in. (from Waters et al.*, 1993)*

	Work duration (h)					
	<1		<2		<8	
Frequency (lifts/min)	$V<75$	$V>75$	$V<75$	$V>75$	$V<75$	$V>75$
0.2	1.00	1.00	0.95	0.95	0.85	0.85
0.5	0.97	0.97	0.92	0.92	0.81	0.81
1	0.94	0.94	0.88	0.88	0.75	0.75
2	0.91	0.91	0.84	0.84	0.65	0.65
3	0.88	0.88	0.79	0.79	0.55	0.55
4	0.84	0.84	0.72	0.72	0.45	0.45
5	0.80	0.80	0.60	0.60	0.35	0.35
6	0.75	0.75	0.50	0.50	0.27	0.27
7	0.70	0.70	0.42	0.42	0.22	0.22
8	0.60	0.60	0.35	0.35	0.18	0.18
9	0.52	0.52	0.30	0.30	0.00	0.15
10	0.45	0.45	0.26	0.26	0.00	0.13
11	0.41	0.41	0.00	0.23	0.00	0.00
12	0.37	0.37	0.00	0.21	0.00	0.00
13	0.00	0.34	0.00	0.00	0.00	0.00
14	0.00	0.31	0.00	0.00	0.00	0.00
15	0.00	0.28	0.00	0.00	0.00	0.00
>15	0.00	0.00	0.00	0.00	0.00	0.00

75 cm = 30 in.

sagittal plane. Measure at the origin and at the destination of the lift (degrees).

The calculations are performed twice – for the point of origin and the point of destination. If the point of destination does not involve *controlled* lifting, for example when the lifter drops the object in place, the latter calculation is excluded.

5.4.1.1 Example: Loading Punch Press Stock

The normal job of a punch press operator is to feed small parts into a press and remove them (Putz-Anderson and Waters, 1991). Once per shift the operator is required to load a heavy reel of supply stock from the floor to the machine (i.e. to a height of 160 cm), as illustrated in Figure 5.5. The reel is 75 cm in diameter and weighs 20 kg. Assume that the operator lifts the reel in the sagittal plane (in front of the body) as shown, and that to load the reel the operator must exercise significant control at the destination of the lift.

Solution:

H Origin	*H* Destination	*V* Origin	*V* Destination	*F*
57.5 cm	57.5 cm	38 cm	160 cm	<0.2

For the origin:

$$\begin{aligned} RWL &= 23 \times HM \times VM \times DM \times AM \times FM \times CM \\ &= 23\times(25/57.5)\times(1-0.003|38-75|)\times(0.82+4.5/122)\times1.0\times1.0\times1.0 \\ &= 7.6\,kg \end{aligned}$$

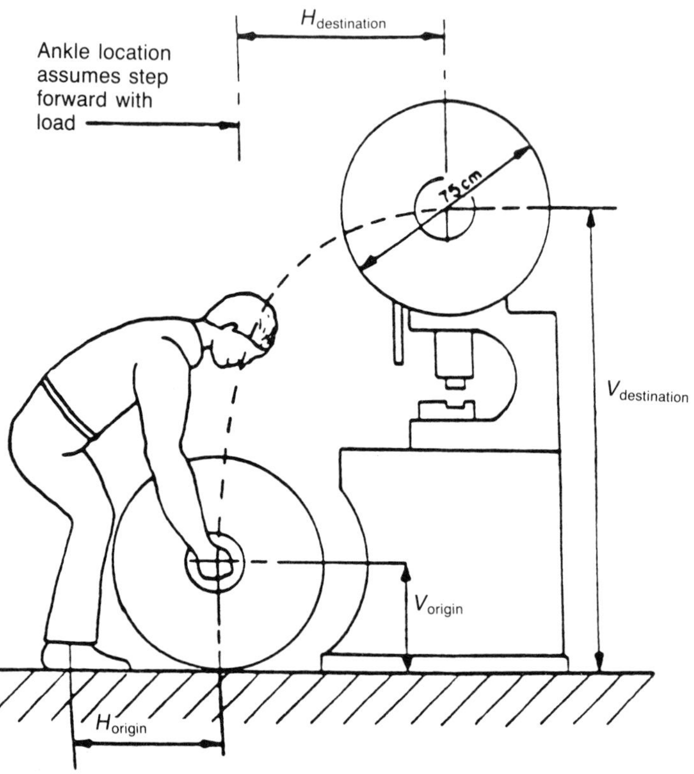

Figure 5.5 Calculation of NIOSH lifting limits: loading punch-press stock

For the destination:

RWL = 23x(25/57.5)x(1 – 0.003|160 – 75|)x(0.82 + 4.5/122)x1.0x1.0x1.0
= 6.4 kg

Because the operator must exercise significant control to load the reel, the calculation for the destination is required. The most protective of the two RWL values is used to estimate the job demands: the RWL for the destination is 6.4 kg, which is smaller than the RWL at the origin (7.6 kg). According to the lifting index formulated below, RWL may be multiplied by a factor of 3 which brings the load to about 20 kg.

5.4.1.2 Example: Product Packaging

In this example products arrive via a conveyor at a rate of 1 per minute (Putz-Anderson and Waters, 1991). The worker packages the product in a cardboard box and then slides the packaged box to a conveyor behind table B as illustrated in Figure 5.6. The product weighs 7 kg (16 lb), and the job is performed for an 8-hour shift. For this example, assume that significant control of the object is not required at the destination. Workers twist their body to pick up the product. Furthermore, assume that workers can flex the fingers to the desired 90° angle to grasp the container. The job is performed for a normal 8-hour shift, including regular rest allowance breaks.

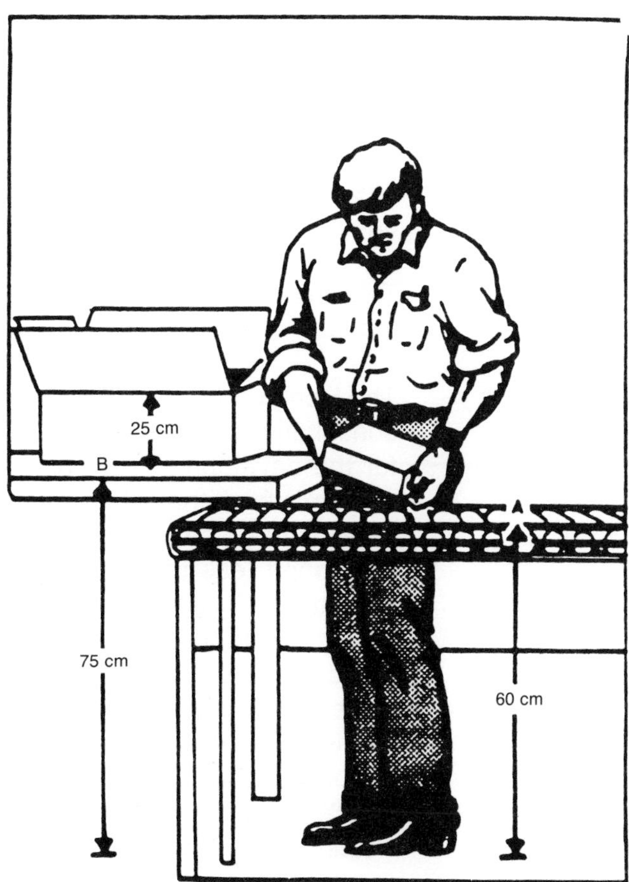

Figure 5.6 Calculation of NIOSH lifting limits: product packaging

Solution: the task data are:

H Origin	*H* Destination	*V* Origin	*V* Destination	*F*	Asymmetry Origin	Asymmetry Destination	Coupling
35 cm	33 cm	60 cm	100 cm	1/min	90°	0°	Fair

Since the worker can grasp the object with the fingers flexed at 90°, the couplings are classified as 'fair' (Waters *et al.*, 1993). In this example, the RWL is only computed at the origin of the lift, since significant control is not required at the destination.

For the origin:

$$\begin{aligned} RWL &= 23 \times HM \times VM \times DM \times AM \times FM \times CM \\ &= 23 \times (25/H) \times (1 - 0.003|V - 75|) \times (0.82 + 4.5/D) \times (1 - 0.0032A) \times 0.75 \times 0.95 \\ &= 23 \times (25/35) \times (1 - 0.003|60 - 75|) \times (0.82 + 4.5/40) \times (1 - 0.0032(90)) \times 0.75 \times 0.95 \\ &= 7.4 \text{ kg} \end{aligned}$$

Thus the recommended weight limit is 7.4 kg, which is about the same as the actual product weight of 7 kg.

5.4.1.3 *Lifting Index*

The lifting index (LI) provides a simple estimate of the hazard of an overexertion injury for a manual lifting job:

$$LI = \frac{\textit{Load of weight}}{\textit{Recommended weight limit}} = \frac{L}{RWL}$$

where L is the weight of the object lifted (lb or kg).

In their discussion of the lifting index, the NIOSH conceded that lifts are often greater than RWL (Waters *et al.*, 1993). Several experts agree that the lifting index should not exceed 3.0, because many individuals would be at a great risk.

5.4.2 Guidelines for the European Community

The Council of the European Communities has also formulated qualitative requirements for manual handling of loads 'where there is a risk of back injury to workers' (EC Council Directive L156, 1990). This directive mandates employers to organize workstations to make manual handling a safe activity. Several factors are listed in Table 5.6.

5.4.3 Guidelines for Manual Lifting in the UK

The Health and Safety Commission (1991) in the UK has developed consultative guidelines for materials handling (Table 5.7). The criteria for the development of the guidelines was to consider a boundary 'beyond which the risk of injury is sufficiently great to warrant a more detailed assessment of the work system'. The guidelines are for lifts performed less than once per minute 'under relatively favourable conditions'. This implies a stable load which is easy to grasp and an upright work posture with a non-twisted trunk. Under such circumstances the guideline figures are assumed to provide reasonable protection to nearly all men and between one-half and two-thirds of women. There are also correction factors for stooping and twisting the body. For example, for 90° stooping the weight should be reduced by 50%, and for 90° twisting it should be reduced by 20%. One major advantage of these guidelines is that they are very easy to use.

5.5 Materials Handling Aids

In an industrial facility there are many different needs for materials handling: transportation of goods to and from the facility; unloading

Table 5.6 Work, environment and personal factors to be considered in workstation organization (EC Council Directive LI56, 1990)

1. *Characteristics of the load*
   The manual handling of a load may present a risk particularly of back injury if it is:
   - too heavy or too large
   - unwieldy or difficult to grasp
   - unstable or has contents likely to shift
   - positioned in a manner requiring it to be held or manipulated at a distance from the trunk, or with a bending or twisting of the trunk
   - likely, because of its contents and/or consistency, to result in injury to workers, particularly in the event of a collision
2. *Physical effort required*
   A physical effort may present a risk particularly of back injury if it is:
   - too strenuous
   - only achieved by a twisting movement of the trunk
   - likely to result in a sudden movement of the load
   - made with the body in an unstable posture
3. *Characteristics of the working environment*
   The characteristics of the work environment may increase a risk particularly of back injury if:
   - there is not enough room, in particular vertically, to carry out the activity
   - the floor is uneven, thus presenting tripping hazards, or is slippery in relation to the worker's footwear
   - the place of work or the working environment presents the handling of loads at a safe height or with good posture by the worker
   - there are variations in the level of the floor or the working surface, requiring the load to be manipulated on different levels
   - the floor or foot rest is unstable
   - the temperature, humidity or ventilation is unsuitable
4. *Requirements of the activity*
   The activity may present a risk particularly of back injury if it entails one or more of the following requirements:
   - overfrequent or overprolonged physical effort involving in particular the spine
   - an insufficient bodily rest or recovery period
   - excessive lifting, lowering or carrying distances
   - a rate of work imposed by a process which cannot be altered by the worker
5. *Individual risk factors*
   The worker may be at risk if he/she:
   - is physically unsuited to carry out the task in question
   - is wearing unsuitable clothing, footwear or other personal effects
   - does not have adequate or appropriate knowledge or training

of materials at the receiving department; transportation of materials to workstations until the product has been assembled, tested and inspected; and transportation of the product to packaging and to a warehouse for final distribution to customers. In addition to these *primary transportation needs* there are also *secondary transportation requirements*, e.g. removal of waste products and housekeeping. Transportation and materials handling in manufacturing constitute a

Table 5.7 Guidelines for lifting according to the Health and Safety Commission (1991)

Height	Less than half arm's length (kg)	Between half arm's length and full arm's length (kg)
Below knee height	10	5
Knee height–knuckle height	20	10
Knuckle height–elbow height	25	15
Elbow height–shoulder height	20	10
Shoulder height–full length	10	5

major expense. We therefore have a dual interest in designing an effective materials handling system:

- To reduce manufacturing costs.
- To reduce ergonomic costs and injuries.

The planning for materials handling and smooth transportation should start at the product design stage (Grossmith, 1992). One important aspect of product design is 'design for ease of handling and transportability'. Thus a product could have a smooth bottom, which makes it easier to transport on conveyor belts. The product can also be equipped with handholds (permanent or temporary) to simplify manual lifting.

The product design is also important because, by virtue of the design features, certain manufacturing processes will become necessary. The process equipment may be available in only one part of the plant, and a 'transportation need' is then created. It may be possible to move the process equipment, so that it is practical for the manufacture of a specific product. However, there is usually a mix of products, and expensive process machinery must be used for many different products. Such issues are then important for the design and layout of a manufacturing facility.

The purpose of just in time (JIT) manufacturing is to structure the transportation activities. According to this philosophy, smaller quantities of parts are delivered to a manufacturing plant and then distributed to workstations, just in time for processing and assembly. The JIT philosophy has an interesting effect in that the need for storage of parts and products is reduced. Therefore, the manufacturing plant can be made smaller, and the cost of buying the land is also reduced (in Japan the cost of land is very high which favours JIT).

5.5.1 Materials Handling Devices

Table 5.8 presents a list of materials handling devices. There are many possible usages of materials handling devices in receiving, at workstations, between workstations, testing, packaging and warehousing. The usage depends entirely on the application and the task at hand. We cannot suggest any fixed formula, it depends on the creativity of the designer. Several of the aids are illustrated in Figure 5.7.

Some devices are used for horizontal transportation in the plant and some of them for vertical transportation. From the ergonomics points of view it is particularly important to minimize vertical transportation, particularly if manual lifting is involved. (Don't put it on the floor, and you won't have to pick it up again.) It is difficult to avoid horizontal

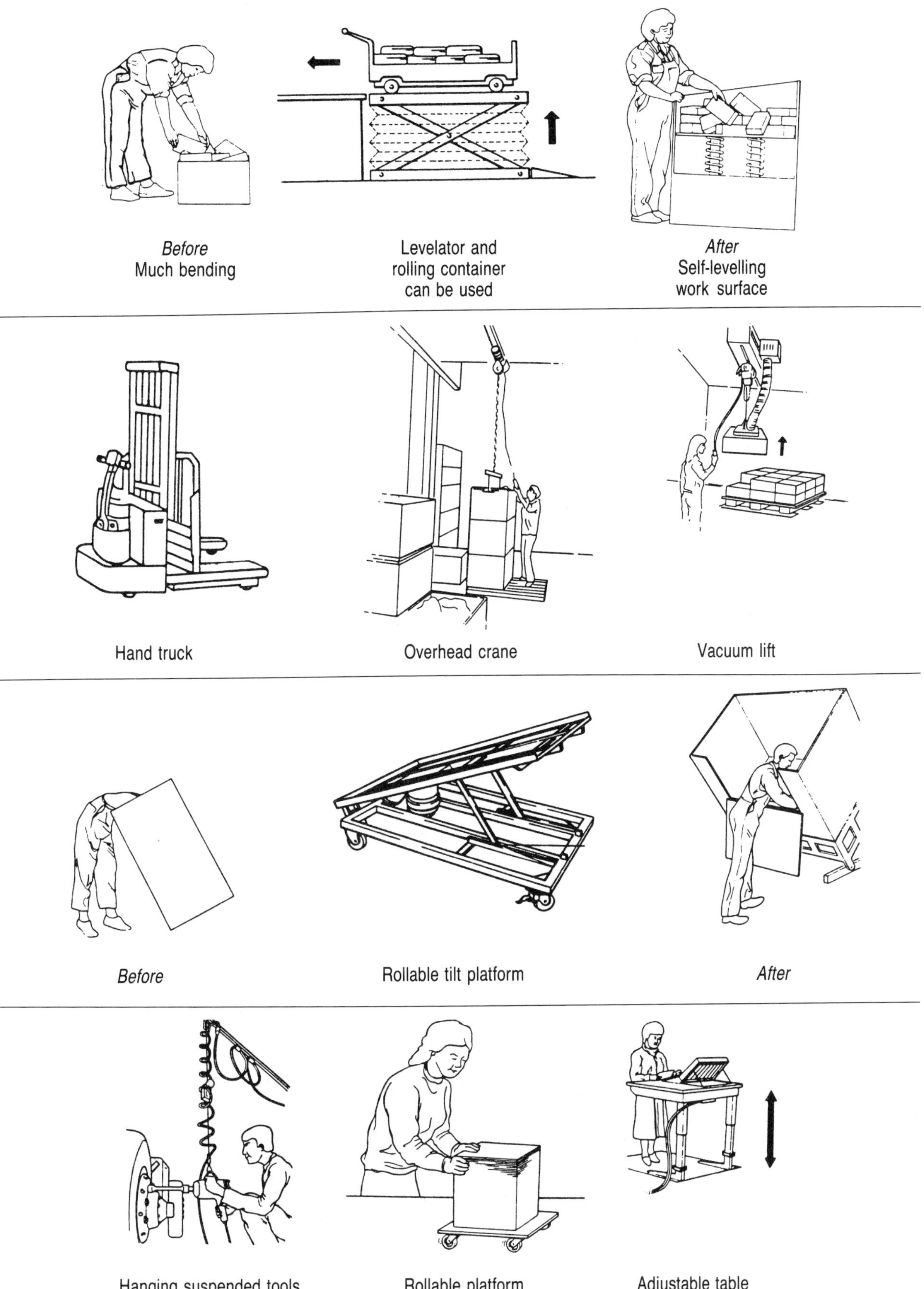

Figure 5.7 Illustration of some aids for lifting and materials handling (courtesy of Swedish Work Environment Fund, 1985)

Table 5.8 A list of manual materials handling devices and their possible use in manufacturing

	Horizontal (H) or vertical (V) transportation	Receiving	At workstation	Between workstations	Testing	Packaging	Warehousing
Conveyors	H	X	X	X	X	X	X
Snake conveyors (easily moveable)	H	X					X
Ball transfer table	H			X	X		
Carts	H		X	X	X		
Carousels	H		X	X	X		
Turntables	H			X	X		
Cranes	V	X	X	X	X	X	X
Hand trucks/walking	H	X		X			X
Forklift trucks	H, V	X	X			X	X
Gravity feed conveyors/slides	H, V				X	X	
Automatic storage/retrieval	V, H	X					X
Stackers	V, H	X					X
Lift/tilt table	V		X		X	X	
Levelators	V		X		X	X	
Scissor table	V		X		X	X	
Vacuum lifting devices	V	X	X			X	X
Self-levelling table	V		X		X		
Adjustable table	V		X		X		
Overhead balancer	V		X		X		

transportation in the plant, although one can try to minimize the transportation distance by optimizing the layout of the facilities. For horizontal transportation, conveyors have generic applicability and can be used for all the different manufacturing stages, including storage in the warehouse. Carts and carousels are also fairly generic, and can be used at several sequential processes. A cart can be used as a moveable workstation that is passed down the line. It can be designed so that an operator can work conveniently at the cart.

Horizontal transportation is continuous and connects the different manufacturing functions. Vertical transportation is mostly local and discrete and does not connect the different systems. It would be an interesting maxim for the design of a plant to 'minimize the vertical movement'. This can be done, for example, by removing the top and the bottom shelfs in storage. For JIT, with its minimal requirement for storage, this is not an unrealistic scenario.

Three of the vertical devices listed in Table 5.8 are automatic: self-levelling tables, gravity feed conveyors, and overhead balancers. These devices are particularly interesting because they do not require any action by the worker.

5.5.2 Exercise: Materials Handling Devices

Please discuss the materials handling devices in Table 5.20. If possible, make a study visit to a local manufacturing company. Make a map of the facilities and indicate which handling devices are used. Then, propose a redesign of the materials handling and transportation. Discuss how the devices listed in Table 5.8 can be used to rationalize transportation and improve ergonomics.

5.6 Recommended Reading

The reader is referred to the books by Chaffin and Anderson (1991), Grandjean (1988) and Kroemer *et al.* (1994).

Chapter 6

Choice of Work Posture: Standing, Sitting, or Sit–Standing

Engineers who design production processes take on a great responsibility. In designing products and processes one must consider how the workstation will be laid out and what type of work posture is convenient for the job. Many engineers, however, focus on the engineering aspects and the design for the worker is done as an afterthought, using fairly simplistic methods. In this chapter we propose criteria for when sitting or standing is appropriate. We also provide measures for appropriate work height.

6.1 Examples of Work Postures

One conventional engineering solution is the prescription of an 'industrial height' workstation with a 92 cm (36 in.) high work table. This can accommodate both sitting and standing operators. The working height for the standing operator is about 92 cm (36 in.), and a sitting operator can use a high chair with a footrest or a ring support. Such flexibility in a workplace is indeed desirable and Figure 6.1 illustrates how flexibility for sitting or standing can be advantageous for many tasks.

However, sometimes the choice of a conventional industrial height workstation creates problems. It is not an appropriate design solution for dedicated seated tasks, which is illustrated by the microscope workstation shown in Figure 6.1(C). Working with a microscope is a seated task. This job requires a very precise and static work posture. There is no reason to consider a standing work posture, and hence a regular table should have been used. This would have one important benefit in that the operator can put his or her feet on the floor, which improves comfort.

Engineers are sometimes short-sighted in their concentration on the technical aspects of the problem. Figure 6.2 illustrated four operators working with very expensive machinery. From these examples it may seem that the greater the technological challenge, the greater is the likelihood that the human element will be forgotten.

Figure 6.2(A) shows an operator working with a scanning microscope. This was a low magnification microscope, with a large exit pupil and it was fairly easy to look through. (High magnification microscopes have a small exit pupil and cannot tolerate any deviation in eye position.) The first obvious problem was that the operator was standing. The second problem was that the scanning microscope moved back and forth while scanning. Therefore the operator had to move back and forth while looking into the microscope – a very demanding task. Due to the bent-over, standing posture, only short operators could perform the task. Rather than having the microscope move, the inspected elements should have moved. Another (less

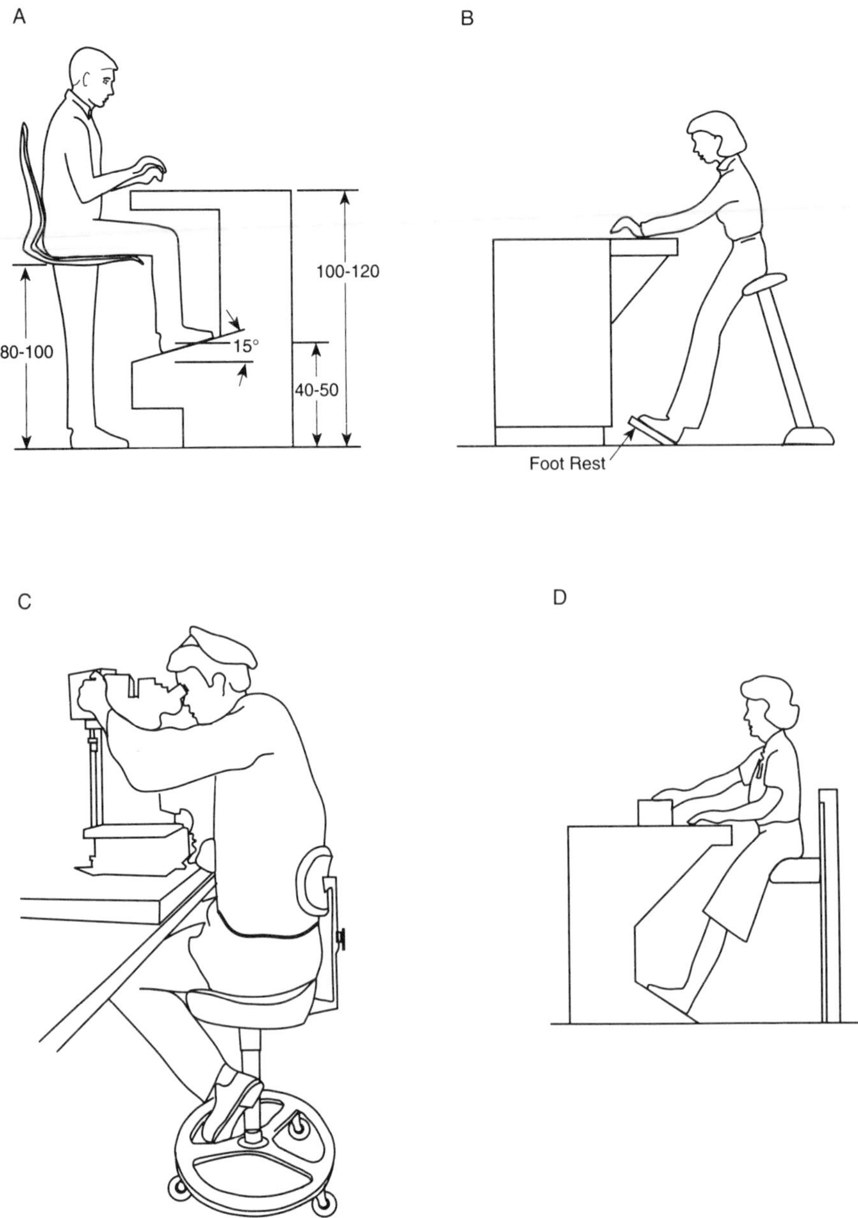

Figure 6.1 (A) A worktable for alternatively sitting and standing – in this case the table at 110–120 cm is higher than the conventional 92 cm (36 in.) table. (B) and (D) Variations of sit-stand arrangements – the operator is free to alternate between standing and sit-standing. (C) Misapplication of a 92 cm (32 in.) industrial height table – working with a microscope is a dedicated seated task, and a regular height table should be used

desirable) solution is to use a moving chair that is synchronized with the microscope movements.

Figure 6.2(B) illustrates a clean-room process which was supposed to be totally automated. However, this very expensive piece of automation never worked out. Visual inspection tasks needed to be performed by a human operator. The main problem was that there was no leg room for the operator sitting at the machine. An armrest was improvised and put on top of the equipment. The operator could then lean sideways on top of the machine to perform the task. This was a very uncomfortable work posture. The design of this piece of

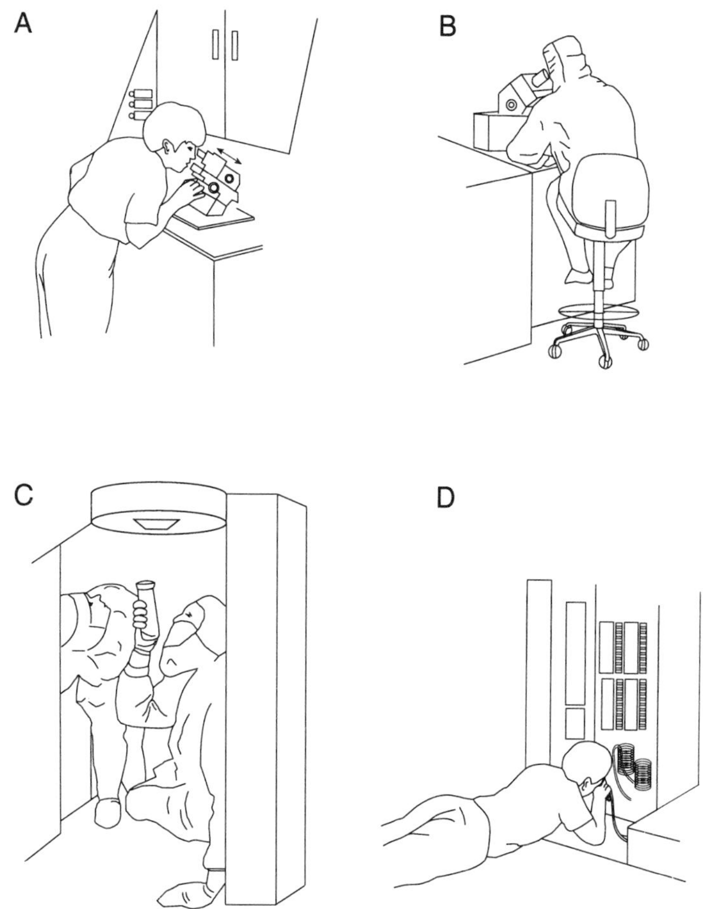

Figure 6.2 (A) The microscope is moving, and unfortunately the operator must move with it. (B) Automated manufacturing equipment – but the automation never worked fully. (C) Routine maintenance with a flashlight. (D) Final assembly of a very expensive piece of equipment

equipment should have made provisions for a manual workstation. Afterwards it became too expensive to rebuild.

Figure 6.2(C) shows two operators in clean-room outfits performing maintenance on process equipment used for manufacturing computer chips. This piece of machinery requires almost constant maintenance, but it was not designed with maintainability in mind (see Chapter 18).

Figure 6.2(D) illustrates an operator lying on the floor completing the final assembly on a piece of electronic equipment which is located inside a steel housing. It was difficult for the operator to reach and to see, and many costly operator errors were reported. As a solution, the company obtained lifting devices to elevate the equipment to a regular working height. The number of operator errors decreased significantly.

It seems inevitable that engineers must expand their responsibilities to consider the consequences of process design for the types of activity that are created. Workstations with ergonomic problems are unproductive, provoke human errors and create costly quality defects. Sometimes the functionality and engineering perfection of a technical system must be compromised to make it more human. Meister (1971) pointed out that engineers are not unwilling to consider the human operator, but they clearly place a higher priority on engineering problems. This view has to change. A design engineer must bring to the task all the relevant tools and skills to solve problems – technological as well as organizational and individual. Engineering design is a systems problem. Technological solutions can only be considered in conjunction with their environment.

6.2 Identifying Poor Postures

The types of posture that people assume at work can often lead to pain in specific parts of the body. Van Wely (1970) reported that there are certain common complaints for different work postures. Table 6.1 summarizes his observations. This table represents an oversimplification. People usually move around, and it is not easy to characterize a job in terms of a single posture. Nonetheless, the list in Table 6.1 is useful as a checklist for inspections of industrial workstations. For example, if one were to observe an operator who sits with his or her elbows on a high surface, it is a reasonable hypotheses that if the operator has any problems they would be in the upper back or lower neck. If the operator indeed voices such complaints, then our hypothesis has been confirmed, and one would reasonably take measures to improve the work posture by lowering the work height. Similarly, if an operator sits with the head bent back the common complaint is neck pain. If someone is assuming a cramped work posture, without any possibility of moving around, then the muscles involved may hurt.

A joint that is in an extreme position, either fully flexed or fully extended, may develop biomechanical problems. Rather, joints should be at a mid-range position. For example, arms should not be fully extended or flexed. A few examples are given in Figure 6.3.

The recommendations for work posture and the discussions about biomechanical problems are traditional in ergonomics. Yet there are problems that require basic research, as is evident from the following example.

6.2.1 Example: Sitting in India

Professor R. S. Sen from the University of Calcutta in India explained that industrial workers in India often sit hunched directly on the floor without a chair, or they may sometimes sit on a brick (Sen, 1989). They develop motion patterns that can be very different from industrial workers in Western countries. Sometimes they swing their knees back and forth to manipulate items, at the same time as they work with their hands. Although their knees are flexed in an extreme position,

Table 6.1 Work postures and related complaints (Van Wely, 1970)

Posture	Complaint
Standing	Feet, lower back
Sitting without lower back support	Lower back
Sitting without back support	Central back
Sitting without proper foot support	Knees, legs, lower back
Sitting with elbows on a high surface	Upper back, lower neck
Unsupported arms or arms reaching up	Shoulders, upper arms
Head bent back	Neck
Trunk bent forward	Lower back, central back
Cramped position	Muscles involved
Joint in extreme position	Joints involved

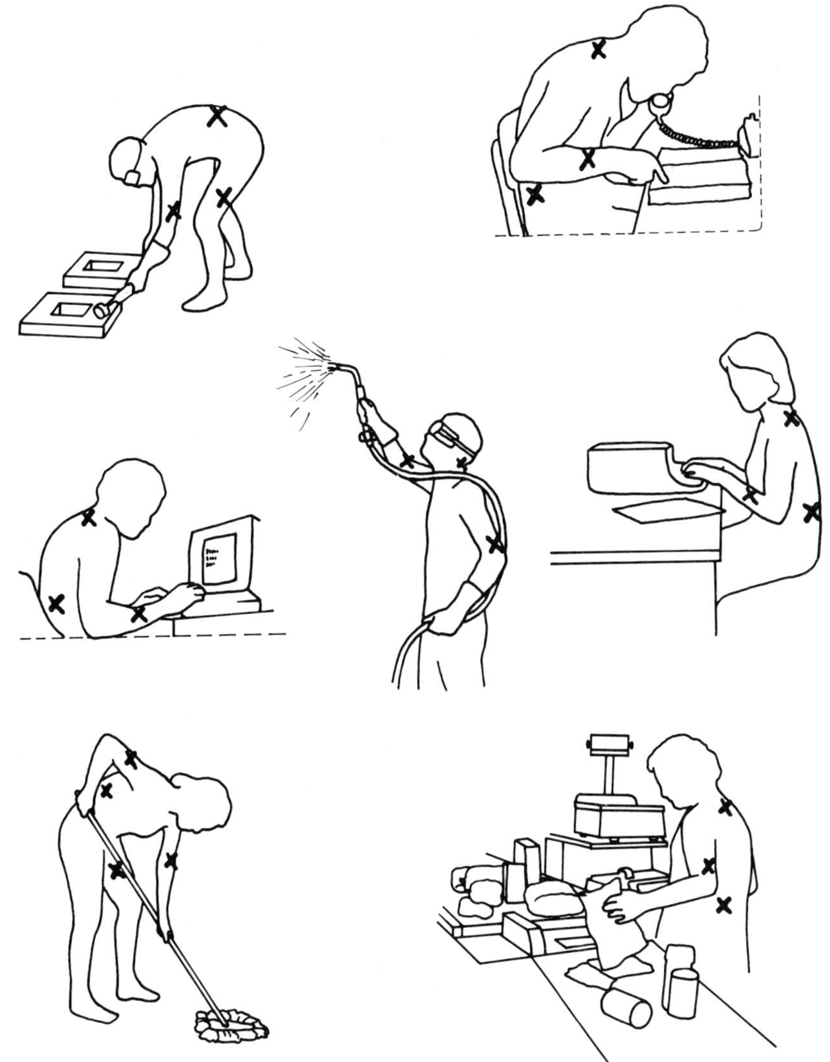

Figure 6.3 Examples of work postures where there are problems with: extreme joint angle, large muscular force, high degree of repetition or high contact pressure (from Webb, 1982)

Professor Sen asserted that these workers do not have any problems with their knee joints. The reason may be that they have been hunch-sitting for their entire lives, and this is a common sitting posture at home or in social gatherings. Professor Sen's statement was surprising, since hunch-sitting violates the principle of keeping the joints in a mid-range position. It seems obvious that more basic research is necessary to analyse this controversy.

6.3 Sitting, Standing or Sit-Standing

We restrict our discussion here to the choice of common work postures. There are additional recommendations in this book, particularly in Chapter 7.

Depending upon the type of task, it may be advantageous for an operator to stand, sit, or sit-stand (Eastman Kodak, 1983; Michel and Helander, 1994).

- If there is frequent handling and lifting of heavy objects it is preferable to stand up. However, sit-standing may be an option (see Table 6.2).

Table 6.2 Preferred work posture for different tasks

Type of task	Preferred work posture	
	First choice	Second choice
Lifting more than 5 kg (11 lb)	Standing	Sit-standing
Work below elbow height (e.g. packaging or assembly)	Standing	Sit-standing
Extended horizontal reaching	Standing	Sit-standing
Light assembly with repetitive movements	Sitting	Sit-standing
Fine manipulation and precision tasks	Sitting	Sit-standing
Visual inspection and monitoring	Sitting	Sit-standing
Frequent moving around	Sit-standing	Standing

- For packaging, or other tasks where objects must be moved vertically below the elbow height, it is preferable to stand or sit-stand. A sitting posture would not be feasible since the hands are reaching downwards and the table cannot be put at a sufficiently low level without interfering with the operator's legs.
- If the task requires extended reaching it is sometimes preferable to stand or sit-stand, as the operator can then reach further.
- Light assembly with repetitive movements is a common task in industry, and sitting is preferable. A table is necessary to organize part bins, fixtures and incorporate work aides and supports to relieve local body fatigue due to repetitive movements.
- For fine manipulation and precision tasks the operator usually wants to support the underarms. Sitting is definitely preferred.
- Visual inspection and monitoring is best done sitting. The sitting work posture makes it possible to focus one's attention better than if standing.
- If the work task involves a variety of subtasks and also frequent moving around, it may be preferable to sit-stand, since the operator does not then have to get in and out of the chair.

The recommendations in Table 6.2 represent a simplification, since there may be other task characteristics that could influence work posture. The recommendations should therefore be used as a first approximation in understanding what the main options are. As we have discussed elsewhere, a task analysis is helpful in understanding the advantages and disadvantages with various design parameters, and how they trade off.

For most of the tasks in Table 6.2 the sit-standing posture is the second choice. This arrangement has become increasingly common in industry during the last 10 years. Sit-standing is convenient for many tasks, and there are biomechanical advantages since the pressure on the spine and the lower back is about 30% lower for sit-standing (and standing) as compared with sitting (Andersson and Örtengren, 1974).

6.4 Hand Height and Determination of Table Height

There are standard recommendations in the ergonomic literature for table (work surface) height for seated and standing workplaces (Ayoub, 1973; Kroemer *et al*., 1994). Figure 6.4 illustrates that for a tall product, the work table must be put at a lower height than for a flat product

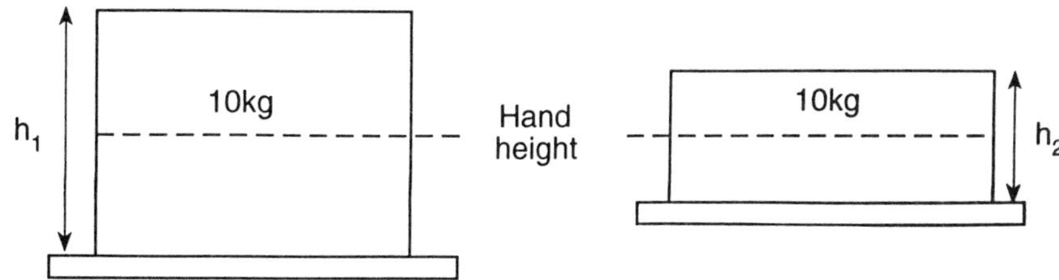

Figure 6.4 The table height is partly determined by the height of the product and the hand height (it is assumed that the products are manipulated or held at the intermediate height)

to arrive at a suitable hand position (Eastman Kodak Co., 1983).

The most advantageous hand position depends on the type of task. For heavy work, it is most convenient to hold the hands about 15 cm (6 in.) below elbow height. The arms and the body can then exercise a greater leverage to perform the heavy task more efficiently. For light assembly work the preferred hand height is about 5 cm (2 in.) below elbow height.

Typing is often performed with the hands about 3 cm (1 in.) above elbow height. For precision work with supported elbows and/or supported underarms, the hand height should be about 8 cm (3 in.) above elbow height. It is easier to perform precision work with the hands and underarms supported. Another reason is that precision work involves small parts and fine details which can be viewed more easily if the objects are closer to the eyes (at about reading distance).

There are individual preferences in work posture. In typing, for example, some individuals may prefer to work with horizontal underarms, but others prefer to raise the keyboard to a higher level. Therefore, the values listed in Table 6.3 are intended as guidelines rather than absolute recommendations. Individuals indeed have different preferences which, combined with anthropometric requirements of a 5th to 95th percentile design, result in a fairly wide range of values. To arrive at suitable values for table height or work bench height from Table 6.3 the handling height of the product must be deducted.

Table 6.3 Measures (cm) of preferred hand height over the floor

		Preferred hand height over floor* (cm)			
	Hand height =	Standing (5th–95th)		Sitting† (5th–95th)	
Type of task	Elbow height ±	Male	Female	Male	Female
Heavy lifting	−15 (Range: −20 to −10)	91–110	85–110	Not recommended	
Light assembly	−5 (Range: −10 to 0)	101–120	95–110	59–79	55–73
Typing	+3 (Range: 0 to +6)	109–128	103–118	67–87	63–81
Precision work	+8 (Range: +5 to +10)	Not recommended		72–92	68–91

*The range for females and males are 5th to 95th percentile (see Table 3.2) and were obtained by deducting or adding the value for hand height. Shoe height of 3 cm is included. 1 in. = 2.54 cm.

†These measures were derived by adding popliteal height, sitting elbow height and shoe height. Note that a height-adjustable chair is assumed, with: Chair seat height = Popliteal height + Shoe height.

6.4.1 Example 1

In this industrial task, 25 kg boxes are transported on a conveyor belt. The operator must turn them over to label both sides. The boxes are 50 cm high and are handled at half-height (25 cm). Calculate the preferred height of the conveyor belt using a 5th to 95th percentile range for standing male operators.

Solution: from Table 6.3 the range for hand height over the floor is 91–110 cm. Deducting 25 cm gives a range of 66–85 cm for the height of the conveyor.

6.4.2 Example 2

Calculate the range of adjustability for a typing table for female 5th to male 95th percentile operators.

Solution: the range for hand height is 63–87 cm. Assuming a 3 cm high 'home (centre row) row' of the keyboard, the table top height is 60–84 cm (23.5–33.0 in.).

6.4.3 Example 3

In a manufacturing plant, sitting workstations will be used for light assembly. Assuming a female population of workers, and that the hand is held at elbow height minus 5 cm, the hand height above the floor is 55–73 cm. Assume further that the product has a handling height of $H/2$ cm, where H is the product height. What is the maximum product height if the worktable is 3 cm thick?

Solution: the solution to the problem is shown in Figure 6.5:

$$\text{Sitting elbow height} = 5 + \frac{H}{2} + 3 + \text{Thigh clearance (cm)}$$

The 5th percentile female operator has a sitting elbow height of 18.1 cm, which is not enough to accommodate the thigh clearance of 10.6 cm, table thickness of 3 m and a hand height 5 cm below elbow height. In this case $H = -1.0$ cm. Obviously, for small parts assembly this workstation is still acceptable, but if large products are handled

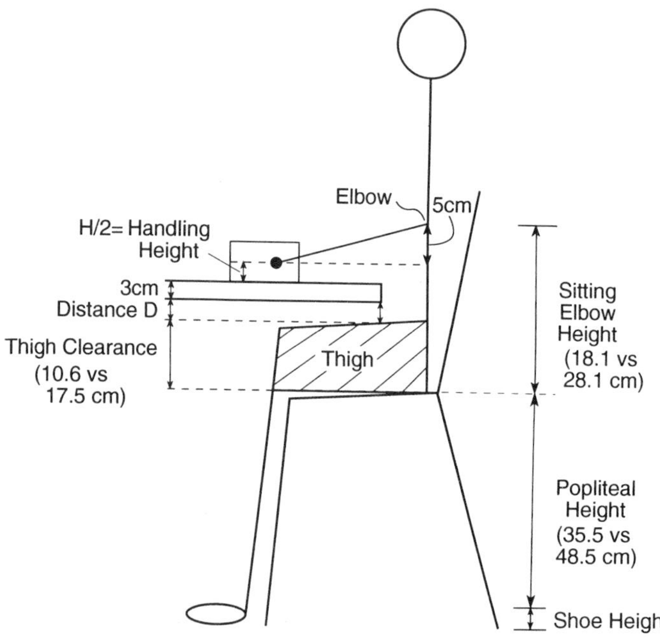

Figure 6.5 Calculation of product height. In the calculations assume that D = 0. The numbers given in parentheses are the 5th and 95th female percentiles

we may want to consider a sit-standing or standing work posture. This does not imply that one would disallow products with greater height at a sitting workstation. Operators can adapt to some extent, for example by gripping the product further down and raising the hands to elbow height. For the 95th percentile female operator, this situation is not so critical because the sitting elbow height is much greater. In this case H = 5.2 cm.

6.5 Work at Conveyors

Conveyors are increasingly used in manufacturing, not only for transportation, but also at assembly lines and for temporary storage, and these systems are often physically connected. At a workstation this arrangement has the advantage that an operator can push items from a moving conveyor to a storage or an assembly line conveyor and is not paced by the line. The operator can thus work faster or slower, as long as the buffer capacity of the storage conveyor is not exceeded (Konz, 1992a).

There is a common belief in industry that the height of the conveyor line must be fixed and consistent throughout a plant. The commonly preferred height is 92 cm (36 in.), which is the same as for industrial standing workstations. This may not always be ideal. Obviously one must avoid downhill and uphill slopes, but there are biomechanical reasons why heights could be different at different locations.

For people working at the conveyors, one should adopt the same rules for determining work height as for regular sitting and standing workstations (see Table 6.3). The purpose is to make the conveyor height convenient for manual work (not for the engineers who design the plant). Thus, the conveyor height should depend on the size of the object that is being handled. For example, if there are large steel drums transported on the conveyor, and if they are handled by workers, then the conveyor height must be very close to the floor to make such handling convenient. Nagamachi and Yamada (1992) demonstrated that the concept of variable conveyor height worked well in a Japanese plant that manufactured air conditioners. The conveyor line was used for assembly and, depending on the height of the work items, the height of the conveyor shifted. They referred to this as a 'Panama Canal Conveyor'. Productivity and quality improved with this design.

If the work along the conveyor is performed sitting, the hand height should be the same as for other sitting workplaces; i.e. for light assembly about 55–79 cm (22–31 in.). There must also be leg room and knee room as for other seated workplaces. In addition, to avoid a bad work posture, the conveyor must be thin so that it can fit in the space between the thighs and underarms. A thick conveyor or a tall fixture will force the operator to raise his or her arms, thereby creating a bad work posture.

Sometimes products on a conveyor line create jams. In order to break up the jams, the conveyor must be accessible from both sides so that two people can work together (Eastman Kodak Co., 1983).

Since conveyor lines can extend throughout an entire plant, it is important to provide crossing points or gates where people and material can be brought through. It should not be necessary to crawl under the conveyor line.

Conveyors can help in manual materials handling at workstations. It should be possible to slide assemblies along the conveyor rather than to lift them. This can be achieved by using special rollers or low friction material which are used to connect a moving and a stationary conveyor at a workstation.

Loading, and especially the unloading, of conveyors present hazards and can result in overexertion and back injuries. Typically unloading is much more demanding and there are three times as many injuries as for loading. This is because the operation is often paced by the movement of the conveyor line, and products typically weigh more when they come off the conveyor line after the assembly (Cohen, 1979).

People working at conveyor belts may develop 'conveyor sickness' (T. G., and R. L., 1975). This may be true not only for moving conveyors but also for other moving objects such as carousel storage units. It seems that if the conveyor speed is greater than 10 m min^{-1} (32 ft min^{-1}) operators can develop nausea and dizziness. This may be particularly common if a person sits sideways to the conveyor, so that the motion is perceived with the peripheral vision.

Chapter 7

Repetitive Motion Injury

Repetitive motion injury, or the more-or-less synonymous term 'cumulative trauma disorder', has become important in ergonomics during the last 10 years. There are many other terms such as 'overuse disorder', 'regional musculoskeletal disorder', 'work-related disorder', 'repetitive distress or strain', 'motion injury', and 'osteoarthrosis'. They are caused by repetitive motions, for example of a hand, and there is a cumulative affect so that a repetitive motion injury may develop over an extended period of time (Putz-Anderson, 1988). In this chapter we use the terms repetitive motion injury (RMI) and cumulative trauma disorder (CTD) synonymously. There are many different types of syndrome; Table 7.1 gives both medical and popular names of some of the more common disorders.

These types of injury have long been recognized in the literature. In the 18th century Ramazzini (1717) described CTD among office clerks, and he believed that these events were caused by repetitive motions of the hand, by constrained body postures, and by excessive mental stress. Liberty Mutual estimated that in 1989 the annual cost in the USA for insurance premiums for CTD cases was $563 million (Webster and Snook, 1994b). This may underestimate the current problem, since during the last 5 or 6 years the reporting of these injuries has increased rapidly.

7.1 Carpal Tunnel Syndrome

The carpal tunnel is an opening in the wrist delimited by the bones of the hand and the carpal tunnel ligament (Figure 7.1). The carpal tunnel is a tight space containing several tendons, some blood vessels and the median nerve. This crowded space is reduced in size even further when the hand or fingers are flexed or extended or bent to

Table 7.1 Some common repetitive motion injuries

Disorder name	Popular names
Carpal tunnel syndrome	Telegraphist's wrists
Cubital tunnel syndrome	Clothes wringing disease
De Quervain's disease	Tennis elbow
Epicondylitis	Golfer's elbow
Ganglion	Bible bump
Shoulder tendonitis	Space invader's wrist
Tendonitis	Slot-machine tendinitis
Tenosynovitis	Pizza palsy
Thoracic outlet syndrome	
Trigger finger	
Ulnar nerve entrapment	

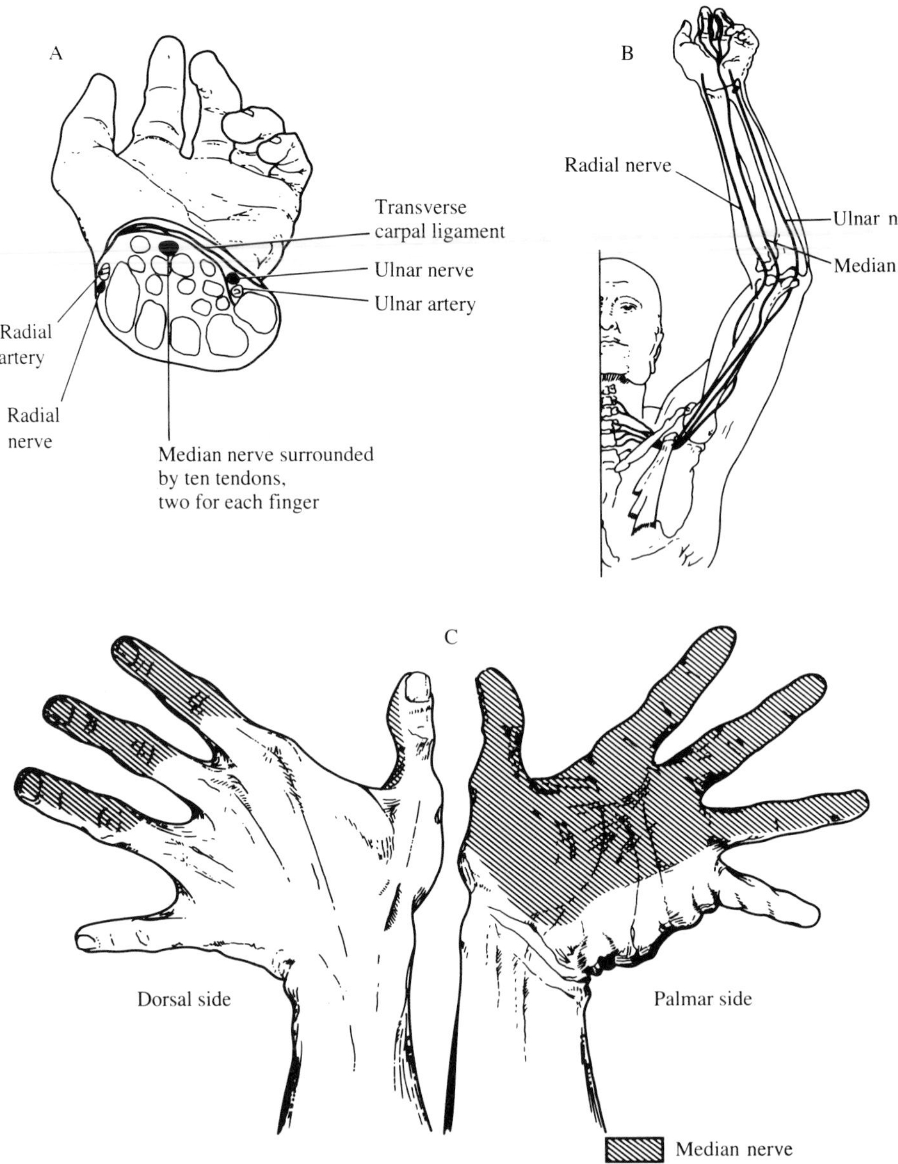

Figure 7.1 (A) Cross-section of the wrist showing the carpel tunnel, which is formed by the five bones on the one side and the transverse carpal ligament on the other. (B) Pathway of the three major nerves that originate in the neck and feed into the arm. (C) Enervation of the hand of the median nerve. The shaded areas indicate where numbness would occur in carpal tunnel syndrome (adapted from Putz-Anderson, 1988)

the side – ulnar deviation and radial deviation. These different postures of the hand are also explained in Figure 7.1(A).

The median nerve enervates the index and middle fingers and the radial side of the ring finger. If there is a swelling inside the carpal tunnel such as would occur if a tendon was inflammed, or if there is external pressure, the median nerve can get squeezed and nerve conduction is no longer efficient. The symptoms of carpal tunnel syndrome are numbness, tingling, pain, and clumsiness of the hand – very much the same as when a foot falls asleep.

Carpal tunnel syndrome has been reported for many occupations in manufacturing (Silverstein *et al.*, 1987). It is particularly significant for meat packers (Brogmus and Marko, 1990) and automobile workers (White and Samuelson, 1990). But it has also been observed among supermarket cashiers (Margolis and Kraus, 1987) and a variety of occupations in manufacturing (Table 7.2).

7.2 Cubital Tunnel Syndrome

This is a compression of the ulnar nerve in the elbow. The ulnar nerve enervates the little finger and the ulnar side of the ring finger, and this is where tingling and numbness will occur. It is believed that cubital tunnel syndrome can be caused by resting the elbow on a hard surface or a sharp edge.

Table 7.2 Repetitive motion injuries reported in manufacturing (adapted from Putz-Andersen, 1988)

Type of job	Disorder	Occupational factors
1. Buffing/grinding	Tenosynovitis Thoracic outlet Carpal tunnel De Quervain's	Repetitive wrist motions, prolonged flexed shoulders, vibration, forceful ulnar deviation, repetitive forearm pronation
2. Punch press operators	Tendinitis of wrist and shoulder	Repetitive forceful wrist extension/flexion, repetitive shoulder abdunction/flexion, forearm supination
3. Overhead assembly (welders, painters, auto repair)	De Quervain's Thoracic outlet Shoulder tendinitis	Repetitive ulnar deviation in pushing controls. Sustained hyperextension of arms. Hands above shoulders
4. Belt conveyor assembly	Tendinitis of shoulder and wrist Carpal tunnel Thoracic outlet	Arms extended, abducted, or flexed more than 60°, repetitive, forceful wrist motions
5. Typing, keypunch, cashier	Tension neck Thoracic outlet Carpel tunnel	Static, restricted posture, arms abducted/flexed, high speed finger movement, palmar base pressure, ulnar deviation
6. Small parts assembly (wiring, bandage wrap)	Tension neck Thoracic outlet Wrist tendinitis Epicondylitis	Prolonged restricted posture, forceful ulnar deviation and thumb pressure, repetitive wrist motion, forceful wrist extension and pronation
7. Bench work (Glass cutters, phone operators)	Ulnar nerve entrapment	Sustained elbow flexion with pressure on ulnar groove
8. Packing	Tendinitis of shoulder wrist Tension neck Carpal tunnel De Quervain's	Prolonged load on shoulders, repetitive wrist motions, overexertion, forceful ulnar deviation
9. Truck driver	Thoracic outlet	Prolonged shoulder abduction and flexion
10. Core making	Tendinitis of the wrist	Prolonged shoulder abduction and flexion. Repetitive wrist motions
11. Stockroom, shipping	Thoracic outlet Shoulder tendinitis	Reaching overhead. Prolonged load on shoulder in unnatural position
12. Material handling	Thoracic outlet Shoulder tendinitis	Carrying heavy load on shoulders

7.3 Tendonitis (or Tendinitis)

This is inflammation of a tendon. The symptoms are pain, burning sensation and swelling. One special case is shoulder tendonitis or bursitis at the rotator cuff (Figure 7.2). Irritation and swelling of the tendon or of the bursa may be caused by continuously keeping the arm elevated or raising the arm (Kroemer *et al.*, 1994).

7.4 Tenosynovitis (or Tendosynovitis)

This is an inflammation of tendons and tendon sheaths. It frequently occurs in the wrist and ankle where tendons cross tight ligaments. The tendon sheath swells which makes it more difficult for the tendon to move back and forth inside the sheaths. Like any inflammation, the symptoms are pain, burning sensation, and swelling.

There are many special cases of tenosynovitis such as De Quervain's disease. This is tenosynovitis of the tendons of the thumb at the wrist. It may occur due to forceful gripping and twisting of the hand, such as using a screwdriver. It has also been called 'clothes wringing disease'. Another special case of tenosynovitis is 'trigger finger', which occurs in the flexor tendons of the finger. The tendon can become nearly locked up so that the movement of the finger is sudden and jerky.

7.5 Thoracic Outlet Syndrome

This is a disorder that results from compression of the three nerves of the arm and the blood vessels (see Figure 7.1(B)). The blood flow to and from the arm is reduced and the arm becomes numb and difficult to move.

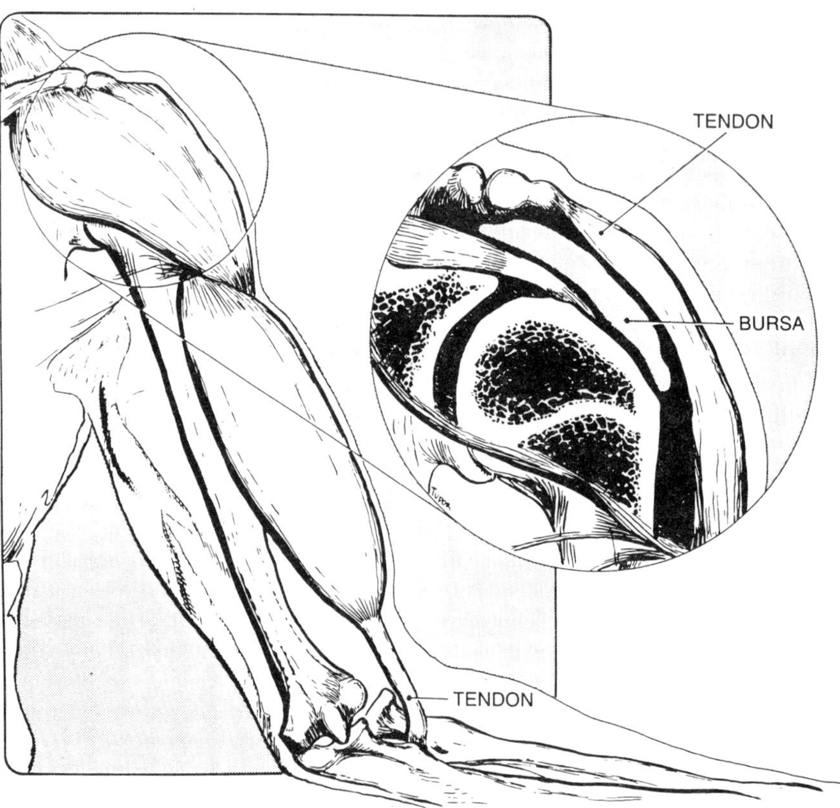

Figure 7.2 A view of the muscle–tendon–bone unit illustrating the relationship between a bursa and a tendon in the shoulder. (From Putz-Anderson, 1988)

7.6 Cause of Repetitive Motion Injury

There are several different factors that may play a part in causing cumulative trauma disorder. For the individual case, it is often impossible to pinpoint a primary cause. One must take a comprehensive look at all the various manual activities that may have contributed to the RMI. It is not just a matter of inappropriate or aggressive work methods, but also what type of activities are performed off work. Leisure activities such as knitting, carpentry and tennis playing will also impact the likelihood of developing RMI. Some of these factors are listed in Table 7.3 (Armstrong and Chaffin, 1979; Eastman Kodak Co., 1986; Putz-Anderson, 1988).

In addition, there may be psychological 'causes' of cumulative trauma disorder. One well-known incidence is the so-called 'RSI epidemic' in Australia. During 1984 the repetitive motion injury rate increased by a factor of 15 (from 50 to 670) among employees of the Australian Telecom. But then the injury rate decreased, and by the beginning of 1987 the injury rate was back to normal (Hadler, 1986; Hocking, 1987). This sudden increase and subsequent drop in injury rate must be attributed to psychological factors. It may have been the case some operators heard that colleagues were having problems and would interpret their own symptoms as being serious manifestations of RMI.

In the last couple of years the RMI rate has increased tremendously in the USA and in Europe, and it would be natural to assume that some of the reported injuries are psychological in nature. But there is also a real problem, and the increased injury rate may be due partly to the situation where it has become accepted in society to report RMI, whereas this was not an accepted work injury in the past. Indeed, Hadler (1989) reported on the types of back injuries reported in Switzerland, Germany and Holland. The legal definitions of back injuries are different in these countries, and as a result different types

Table 7.3 Causes of cumulative trauma disorders (note that many of the listed causes have not been reconfirmed by research, since they are difficult to investigate, and it takes a long time to accumulate epidemiological data)

Inappropriate work methods:
- Repetitive hand movements with high force
- Flexion and extension of hand
- High force pinch grip
- Uncomfortable work postures

Lack of experience of manual work
- New job
- Back from vacation

Inappropriate leisure activities
- Insufficient rest due to working in a second job
- Knitting, playing musical instruments, playing tennis, bowling, home improvements

Pre-existing conditions
- Arthritis, bursitis, other joint pain
- Nerve damage
- Circulatory disorders
- Reduced oestrogen level
- Small hand/wrist size

of back problem are reported. Society norms and acceptance seems greatly to affect the type of occupational injuries that are reported.

Another example is for VDT workers. In the Scandinavian countries complaints of pain in the neck and shoulder are common (Hagberg and Sundelin, 1986), but RMI have been rare (Winkel, 1990). In the USA the situation is different, and carpel tunnel syndrome is frequently reported among VDT operators (National Institute for Occupational Safety and Health, 1992). The shoulders and hands are connected by the three nerves (see Figure 7.1(B)), and there may be a possibility that the aetiology of the injuries is the same, although the manifestation of complaints are different, so as to conform to the local norms.

Whatever reason employees may have (physical or psychological) one must take complaints seriously. There are often simple modifications and additions to workstations that can alleviate some of the problems. For example, VDT operators often ask for a soft wrist

Table 7.4 Guidelines for reducing RMI through product design, process engineering, workstation design and use of appropriate handtools

Guidelines for hand posture

- Watch out for sudden flexion or extension of the hand or fingers
- Avoid extreme ulnar deviation and radial deviation
- Avoid operations that require more than 90° wrist rotation
- Keep forces low during rotation or flexion of the wrist
- For operations that require finger pinches keep the forces below 10 N; this represents 20% of the weaker operators' maximum pinch strength

Guidelines for handtools

- Cylindrical grips should not exceed 5 cm (2 in.) in diameter
- Avoid gripping that spreads the fingers and thumbs apart by more than 6 cm (2.5 in.)
- Use handtools that make it possible to maintain the wrist in a neutral position (see Figure 8.2)

Guidelines for workstation design

- Keep the worksurface low to permit the operator to work with elbows to the side and wrists in a neutral position
- Avoid sharp edges on the work table and part bins that may irritate the wrists when the parts are procured
- Keep reaches within 20 in. from the work surface so that the elbow is not fully extended

Guidelines for process engineering

- Allow machinery to do repetitive tasks and leave variable tasks to human operators
- Provide fixtures that hold parts together during assembly and which can present the assembly task at a convenient angle to the operator
- Minimize time pressure or pacing pressure by allowing operators to work at their own pace

Guidelines for product design

- Minimize the number of screws and fasteners used in the assembly
- Minimize the torque required for screws
- Locate fasteners and screws at 'natural' angles so they are easy for the operator to insert
- Design a product with large parts to permit gripping with fingers and palm instead of pinching

rest, a split keyboard, a lower typing surface, or a footrest. These are inexpensive modifications, and one should not question the utility of such measures.

7.7 Design Guidelines to Minimize Repetitive Motion Injury

Table 7.4 illustrates several engineering guidelines that can be used to minimize RMI. The assumption for presenting these guidelines is that the working environment, the task, and the workstation can be improved or redesigned by using various measures.

Chapter 8

Hand Tool Design

Hand tools have been used since the beginning of mankind, and ergonomics was always a concern. Tools concentrate and deliver power, and aid the human in tasks such as cutting, smashing, scraping and piercing. Various hand tools have been developed since the Stone Age, and the interest in ergonomic design can be traced back in history (Childe, 1944; Braidwood, 1951).

During the last century there has been one important modification – many hand tools are now powered. The forces are greater, and thus the opportunities for injuries are also greater. In this chapter we give some guidelines for designing hand tools. There are several issues. A hand tool must:

- Fit the task.
- Fit the user and hand.
- Not create injuries.

8.1 Fitting the Task

There are two basic grips: the power grip and the precision grip (Figure 8.1). In the power grip, the hand makes a fist with the forefingers on one side and the thumb reaching around. There are three different categories of power grip that are differentiated by the direction of the force: (1) force parallel to the forearm, e.g. a saw; (2) force at an angle to the forearm, e.g. a hammer; and (3) torque about the forearm, e.g. a screwdriver (Konz, 1990).

For precision grips there are two subcategories: (1) the internal precision grip where the tool is held inside the hand, e.g. a table knife; and (2) the external precision grip where the tool is pinched by the thumb against the index finger and middle finger, e.g. a pen.

A hand tool can often be designed in different ways, since there are different ways of exerting power on the tool and the task. An electric screwdriver can have a pistol grip or an inline grip (Figure 8.2), and a surgical knife can be handled with an internal precision grip or an external precision grip. The option chosen should depend on how the task is organized and what is convenient for the operator.

There are many special-purpose hand tools. An accomplished chef has at least a dozen different knives for different purposes. Some of them are handled with a power grip and some with a precision grip and, depending upon the task, they are small or large, flexible or stiff. Likewise, for manufacturing one can design special-purpose hand tools to fit specific tasks. Sometimes it is also possible to combine several hand tools into one. For example, hammers with an extension for pulling nails, which makes it convenient for carpentry. A combination hand tool will save time because the operator can use one tool rather than two.

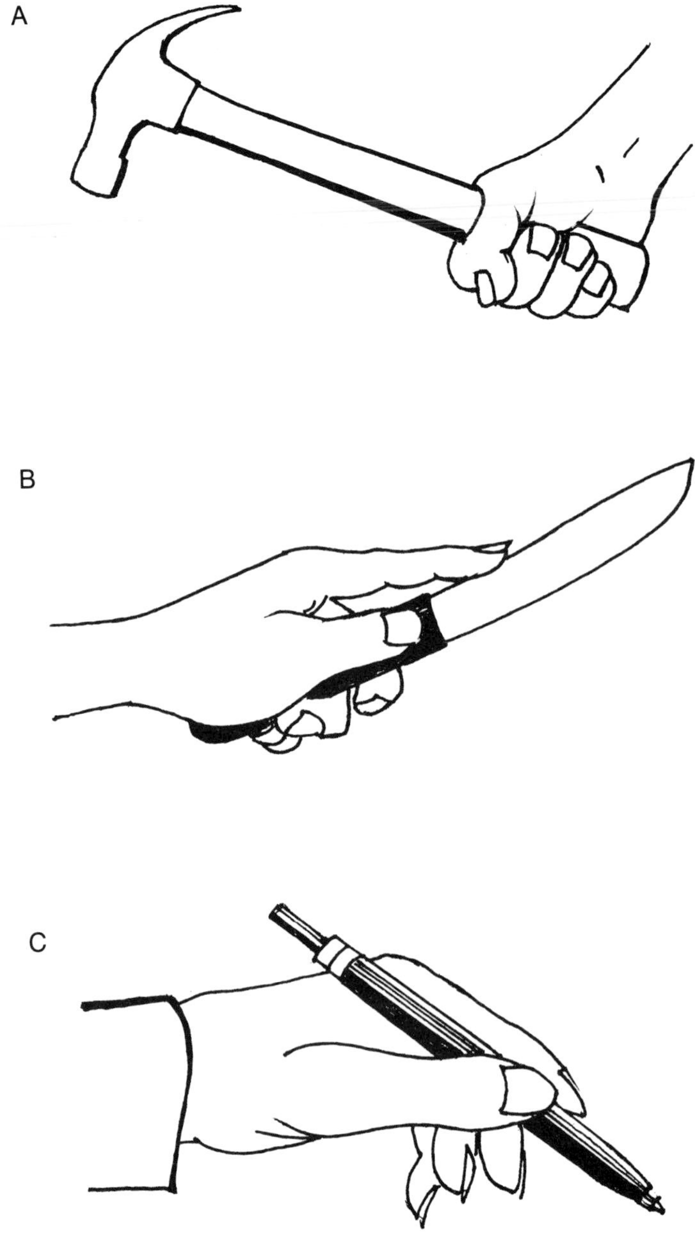

Figure 8.1 (A) Power grip; (B) internal precision grip; (C) external precision grip

8.2 Design for the User

Here we are concerned with the size of the hand and left-handedness versus right-handedness. As demonstrated in Table 3.2, there are few other dimensions of the human body where the differences between the sexes are as great as for the size of the hand (Ducharme, 1973). Typically, the hand circumference for a 5th percentile male is the same as that for a 50th percentile female. Several organizations in the USA, such as General Motors and the US Navy, have a large number of female operators. They now supply hand tools appropriate for the female hand. Figure 8.3 shows the difference in the maximum grip strength for the average male and the average female. The maximum grip force for a female is about half that of a male operator.

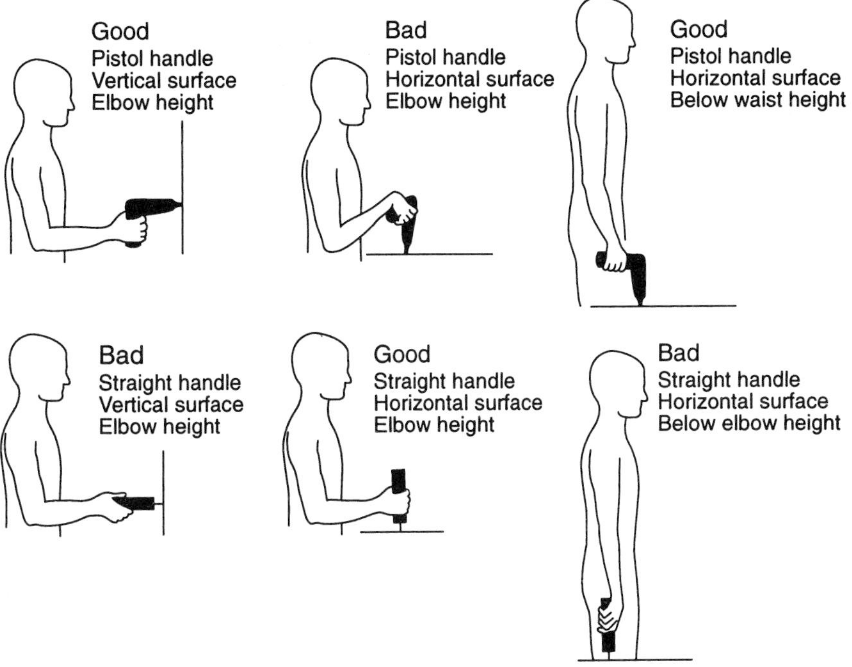

Figure 8.2 A hand tool should be selected so that it is possible to operate with a straight wrist

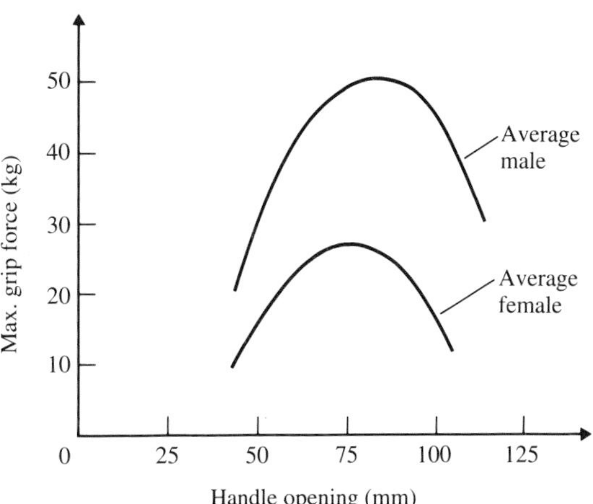

Figure 8.3 Maximum grip strength for male and female employees in electronics manufacturing (adapted from Greenburg and Chaffin, 1977)

Right-handed tools for left-handed users create awkward situations. The left-handed person can try to use the tool with the right hand but his or her dexterity and power is better with the left hand, and productivity will suffer. Sometimes the left hand can grip a right-handed tool, but there may be cut-outs for the fingers which do not fit. Ideally, a hand tool should be designed so it can fit both the left-handed and the right-handed user. Cut-outs for the fingers, for example, should be avoided.

8.3 Prevention of Injuries

There are two major concerns in hand-tool design: injuries due to cumulative trauma disorder (CTD), and vibration-induced injuries. Repetitive usage of hand tools is associated with the development of cumulative trauma disorder, such as carpal tunnel syndrome and tenosynovitis. One common recommendation for preventing CTD is that the movement of the hand should be minimized. Ideally, the hand should be in its neutral straight position, and sometimes handles can be modified to better fit a task. Tichauer's (1966) study of pliers used in a Western Electric plant is a classic example that has inspired many ergonomists. In this case, a plier used for electronic assembly was redesigned. The handles were bent so that it was no longer necessary to bend the wrist to perform the task (Figure 8.4). This design was successful and the incidence of tenosynovitis was reduced significantly among workers. The design motto is: 'it is better to bend metal than to twist arms' (Sanders, 1980).

The hacksaw shown in Figure 8.5 can be designed with different types of handles. In (A) the hand close to the body would be in ulnar deviation, and the hand at the far end would be in dorsiflexion. For

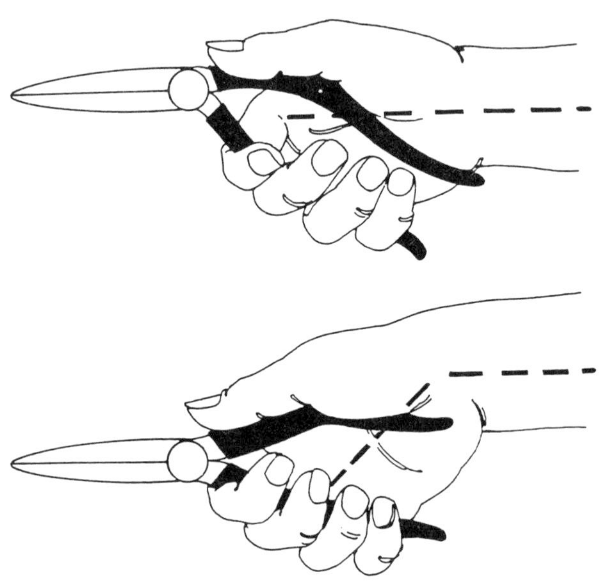

Figure 8.4 The handles of the tool are bent so that the wrist can remain straight

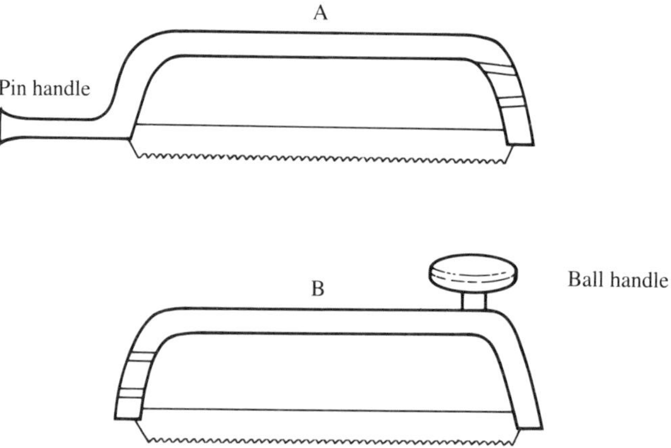

Figure 8.5 Two design options for a hacksaw. Case (B) is clearly better as both hands can operate with straight wrists

case (B), both hands would be perfectly straight and aligned with the tool, so this is a much better design.

John Bennett, an enterprising ergonomist, obtained a patent for the so-called 'Bennett's bend'. This implies using bent (19° ±5°) handles for a variety of different tools (hammers, knives, broom handles and tennis rackets). Investigations by Schoenmarklin and Marras (1989) and Krohn and Konz (1982) verified that a bent hammer handle had the effect of reducing ulnar deviation and did not hamper performance, compared with straight handle hammers. However, there is nothing magic about 19°, and some scepticism would be appropriate. We need further scientific evidence to validate Bennett's bend.

8.4 Segmental Vibration

Hand-tool vibration can cause vibration injuries. There are two common types of vibration injury: Reynaud's disease (or white finger disease) and Dart's disease. Reynaud's disease is caused by hand-tool vibration in the frequency range 50–100 Hz. Examples of such hand tools are pneumatic drills, jackhammers and concrete vibrators. The white fingers are caused by a reduction in blood flow to the hand and to the fingers, which is due to constriction of the smooth muscles of the blood vessels in the hand and fingers. Both the nerves and the blood vessels in the hand are permanently damaged (National Institute for Occupational Safety and Health, 1989).

The reduction in blood flow causes stiffness and numbness of the fingers, and gradual loss of muscle control of the hand. Workers have difficulty in holding, grasping and manipulating items. White finger disease is aggravated by other conditions that cause vasoconstriction of the hand, such as cold weather and smoking. The feeling in the hand is the same as when a foot 'falls asleep', and there are complaints of tingling, numbness and pain.

Table 8.1 Design guidelines for hand-tool design

For precision grip
Grip between thumb and finger
Grip thickness 8–13 mm
Grip length minimum 100 nm
Tool weight maximum 1.75 kg
Trigger activated by distal phalanges of finger(s) with fast-release locking mechanism

For power grip
Grip with entire hand
Grip thickness 50–60 mm
Grip length minimum 125 mm
Grip force maximum 100 m
Grip shape non-cylindrical; preferably triangular with 110 mm periphery
Tool weight maximum 2.3 kg, preferably about 1.2 kg
Trigger activated by thumb with locking mechanism

General guidelines
Grip surface smooth, slightly compressible and non-conductive
Avoid vibration, particuarly in the range of 50–100 Hz
Design handles for use by either hand
Keep the wrist straight in handshake orientation
Tool weight balanced about the grip axis
Eliminate pinching hazards

Dart's disease is less common. This disease is caused by vibration frequencies around 100 Hz. The symptoms are the opposite to those of white finger disease. In Dart's disease, blood pools in the hands, which become blue, swollen and painful.

One way of reducing the transmission of vibration is to use a vibration-attenuating handle. Andersson (1990) used a handle that consisted of a hand grip and a rubber element which acted as a universal joint. This handle effectively reduced transmitted vibration by about 70%. Soft handles such as foam grips do not seem to work. A study by Fellows and Freivalds (1991) demonstrated that grip force was greater when using a foam grip, since the deformation of the foam led subjects to feel as though they were losing control. Due to the increased grip force more vibration energy was transferred to the hand.

8.5 Design Guidelines for Hand Tools

Table 8.1 summarizes several design guidelines for hand tools. The aim of these guidelines is to increase operator comfort, convenience and controllability of hand tools.

Chapter 9

Illumination at Work

A well-designed illumination system is important for industrial productivity and quality, as well as operator performance, comfort and convenience (Hopkinson and Collins, 1970). In this chapter we will explain how to design illumination. Improved illumination is not just a matter of installing more lights, but also how this is done. There are several ways of improving the quality of illumination, for example by using 'indirect lighting'. Such lighting can be important since it reduces the amount of glare. As we will note, older persons are particularly sensitive to glare, which may have a disabling effect on their vision.

We also discuss illumination for visual inspection. Visual inspection can be enhanced by using special-purpose illumination, which makes flaws more visible. Illumination for VDT workstations is discussed in Chapter 10.

9.1 Measurement of Illuminance and Luminance

The distinction between illuminance (also called illumination) and luminance is important. Illuminance is the light falling on a surface. After it has fallen on the surface it is reflected as luminance. Luminance is therefore a measure of light reflected from a surface. Luminance is also used to measure light emitted from a VDT screen. This may be theoretically incorrect, but for practical purposes light from a VDT screen has the same properties as reflected light.

To calculate how much luminance can be generated from a surface one must know how reflective the surface is. This is specified by using measurements of *reflectance*; a number which varies from 0 to 1. It is practically impossible to achieve a perfect reflectance of 1.0: a piece of white paper has a reflectance of about 0.85. A non-reflective black surface has a reflectance of 0.

Measurement units are typically specified in the SI system (metric system). Illuminance is measured in lux and luminance in candela per square metre (cd m^{-2}), also called 'nits'. These are the preferred measurement units (Boyce, 1981b).

In the USA the 'English system' still persists. According to the English system illuminance is measured in foot-candles (fc). One foot-candle equals 10.76 lux, but for practical purposes a conversion factor of 10 is sufficient. Thus 1000 lux illumination which would be appropriate for an industrial workstation corresponds to 100 fc. In the English system luminance is measured in foot-lambert (fL). One foot-lambert is equivalent to 3.4 cd m^{-2} (or 3.4 nits). The measurement units are illustrated in Table 9.1.

There is a simple formula for converting illuminance to luminance.

For the SI system:

$$\text{Luminance (cd m}^{-2}) = \frac{\textit{Illuminance (lux)} \times \textit{Reflectance}}{\pi}$$

Table 9.1 Units for measuring illuminance and luminance (SI units are preferred)

	English	SI
Illuminance (or illumination) – amount of light falling on a surface	1 foot-candle (fc) (or lumen ft^{-2})	= 10 lux (lx) (or lumen m^{-2})
Luminance – amount of light coming from a surface	1 foot-lambert (fL) (or candela ft^{-2})	= 3.4 candela m^2 (cd m^{-2}) (or 3.4 nits)

English system:

$$\text{Luminance (fL)} = \text{Illuminance (fc)} \times \text{Reflectance}$$

9.2 Measurement of Contrast

Contrast is the difference in luminance between two adjacent objects. It is calculated as a *contrast ratio* between the luminances of the two areas A and B:

$$\text{Contrast ratio} = \frac{\textit{Luminance A}}{\textit{Luminance B}}$$

An alternative way of expressing contrast is as *modulation contrast*:

$$\text{Modulation contrast} = \frac{\textit{Luminance}_{max} - \textit{Luminance}_{min}}{\textit{Luminance}_{max} + \textit{Luminance}_{min}}$$

where Luminance_{max} is the greater of the two luminances.

Modulation contrast is less than 1.0. Some experts prefer this expression of contrast since it has properties that better resemble the sensitivity of the human eye (Snyder, 1988).

Both contrast and illuminance are important for visibility. For many items in the working environment contrast is rather high. For example, for black print on white paper the contrast is around 1:40, which provides excellent visibility. However, for characters on a VDT screen a contrast of 1:8 is not unusual, which is somewhat less visible (Shurtleff, 1980).

9.2.1 Example: Contrast Requirements in Manufacturing

In manufacturing assembly visual contrast may be critical. For one particular assembly it was important to distinguish between gold coloured electrodes, copper coloured electrodes and copper oxide. This involved very small details in electronic manufacturing. Operators were looking through a microscope and bonding the copper electrodes to the gold electrodes. Work with microscopes is very demanding, and to relieve the postural strain a TV system with a TV camera and a monitor was brought in. Instead of looking into the microscope the operator could now look at the monitor while still performing the bonding operation manually. However, it turned out that the colour rendering of the TV system was insufficient to distinguish between the rather subtle differences between gold, copper and copper oxide. The TV system had to be removed and the human operator returned to using the microscope.

Very large contrast between large objects can cause *discomfort glare*. For example, the contrast between a window and an adjacent wall is often as large as 100:1. It is a common recommendation not to locate workstations so that the operator will face a bright window. Discomfort glare may cause oscillations of the eye pupils, but people are usually unaware of this phenomenon. Although discomfort glare is harmless, it is nonetheless annoying and discomforting.

For the same reason one should avoid extreme contrasts in the workplace. A common recommendation is that the contrast ratio between the task and large items in the workstation should be less than 10:1 (or greater than 1:10). Some recommendations specify that the contrast between the task and the adjacent surroundings should be less than 3:1 (Illuminating Engineering Society, 1982). However, 3:1 is too restrictive, and 10:1 is more reasonable (Kokoschka and Haubner, 1985). Grandjean (1988) recommended that the maximum luminance ratio within an office should not exceed 40:1.

9.3 Use of a Photometer

Illuminance, luminance, and contrast ratio can be measured with a hand-held photometer. This device is similar to a camera lightmetre, except that it provides a direct readout in lux (or cd m^{-2}). A photometer is colour corrected so that it simulates the human sensitivity to colour. Thus, since the human sensitivity to violet and red (at the opposite ends of the colour spectrum) is less than to green and yellow (at the centre of the colour spectrum), the photometer will produce lower values for violet and red than for green and yellow. Therefore, in determining the luminance one need not be concerned about colour as such, since the photometer will simulate the sensitivity of the human eye.

For example, people often ask which is the best colour for characters on a monochromatic VDT screen: green or yellow or white characters against a black background. The answer is that as long as one uses a photometer to measure the luminance, the results are safe. The character colour that produces the greatest contrast ratio with the monochromatic background gives the best visibility.

Photometers have two different settings: one for measuring illuminance and one for measuring luminance (Figure 9.1). To measure the illuminance that falls on a surface, one must consider contributions from a variety of sources: light sources (luminares), windows and wall

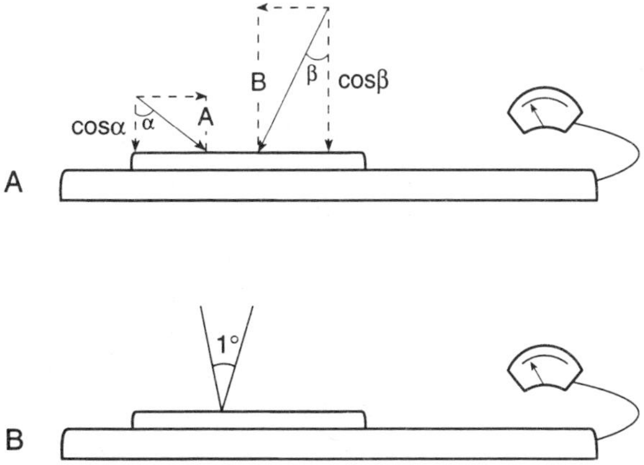

Figure 9.1 Use of a photometer for measuring (A) illuminance and (B) luminance. (A) Wide acceptance angle with a cosine correction. (B) Narrow acceptance angle for spot measurement

reflections. The photometer must have a wide angle of acceptance, and it must be cosine corrected to account for contributions which are not perpendicular to the photocell on the photometer. To measure luminance the photometer must have a narrow angle (e.g. 1°) of acceptance. This enables precise readings of adjacent areas with different reflectance. To measure the contrast ratio between two objects two luminance readings are obtained, and the contrast ratio is calculated.

The contrast ratio between characters and the screen background is important for visibility, but is difficult to measure. The characters are composed of a rectangular array of dots (pixels). To measure the luminance of a pixel, a special photometer with a micro-image slit is required. The procedures are specified in the US Standard ANSI/HFS 100 (Human Factors Society, 1988).

9.4 Recommended Illumination Levels

Many experiments have been performed to determine the appropriate illumination levels for different tasks. Over the years there has been a succession of recommendations, each claiming to provide adequate illumination. The recommended levels, however, continually increase. Current recommended levels are about 5 times greater than the levels recommended 30 years ago for the same tasks (Sanders and McCormick, 1993).

One method for determining the required illumination is based on the laboratory research by Blackwell (Blackwell, 1964, 1967; Blackwell and Blackwell, 1971). The experimental task was to detect the presence of a uniformly luminous disk subtending a visual angle of 4′ (about 1.1 mm at a distance of 1 m). Blackwell found that when the background luminance decreased, the contrast of the just barely visible disk had to be increased to make it just barely visible again. Laboratory studies are not without problems. Blackwell's studies can be criticized for being overly artificial, since there are few real-life situations that resemble his experimental set-up. In addition, subjects in laboratory studies know that they are participating in an experiment and they are usually motivated to perform well.

Field studies also have their problems, since there are many simultaneously independent variables that affect the outcome. These independent variables are usually not manipulated; sometimes they are not even measured. It can therefore be equally difficult to draw conclusions from field studies. As an example of a successful field study, Bennett *et al.* (1977) measured the task completion time for several industrial tasks, while varying the illuminance. The general conclusion was that increasing illumination beyond 1000 lux seems to have limited benefits (Figure 9.2).

The Illuminating Engineering Society (IES) publishes recommended values of illumination (Table 9.2). Depending upon the size of the visual task and the contrast of the task, different levels of illumination are required. These guidelines also take into account the worker's age, the importance of speed and accuracy, and the reflectance of the task background. The upper end of the recommended range in the table should be used to accommodate older workers and the lower values are for younger workers. It is also suggested that local task lighting rather than general ambient illumination be used, particularly if the illumination at the workplace is above 1000 lux.

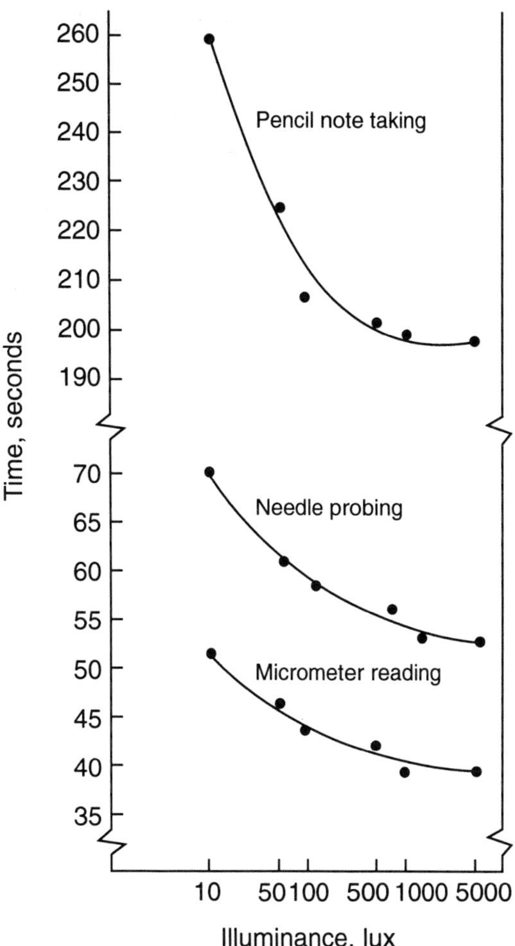

Figure 9.2 Relationship between the amount of illuminance and task completion time (Bennet et al., *1977)*

The IES recommendations prescribe higher levels of illumination than do the European guidelines. This is illustrated below through the comparison with the German DIN standard, and serves to emphasize that different experts have different opinions (Sanders and McCormick, 1993). In many experts' opinion, 5000 lux should be the maximum.

Type of work	*German DIN*	*IES*
Precise assembly work	1000 lux	3000 lux
Very precise machine tool work	1000 lux	7500 lux
General office work	500 lux	750 lux

9.5 The Ageing Eye

For older individuals there are several physical changes that take place in the eye. The most important is the loss of focusing power (accommodation) of the lenses in the eye (Safir, 1980). This is because with increasing age the eye lenses lose some of their elasticity, and therefore cannot bulge or flatten as much as before.

Figure 9.3 illustrates that the average accommodation for a 25-year-old is about 11 diopters, but for a 50-year-old it is only 2 diopters and for a 65-year-old it is 1 diopter. The number of diopters translate into a range of clear vision that is defined by its far point and its near point. Assume, that for the 25-year-old the far point is at infinity. The near

Table 9.2 Illuminance recommended by the IES for industrial tasks (adapted from Kaufman and Christensen, 1984)

Type of task	Range* of illuminance (lux)
Workplaces where visual tasks are only occasionally performed	100–200
Visual tasks of high contrast or large size: printed material, rough bench and machine work, ordinary inspection	200–500
Work at visual display terminals for exended periods of time†	300–500
Visual tasks of medium contrast or small size, e.g. pencilled handwriting, difficult inspection, medium assembly	500–1000
Visual tasks of low contrast or very small size, e.g. handwriting in hard pencil on poor-quality paper, very difficult inspection	1000–2000
Visual tasks of low contrast and very small size over a prolonged period, e.g. fine assembly, highly difficult inspection	2000–5000
Very prolonged and exacting visual tasks, e.g. extra-fine assembly, the most difficult visual inspection	5000–10 000

*The upper values in the range are for individuals aged over 55 years and the lower values are for individuals younger than 40 years.
†This recommendation is from ANSI/HFS 100 (Human Factors Society, 1988).

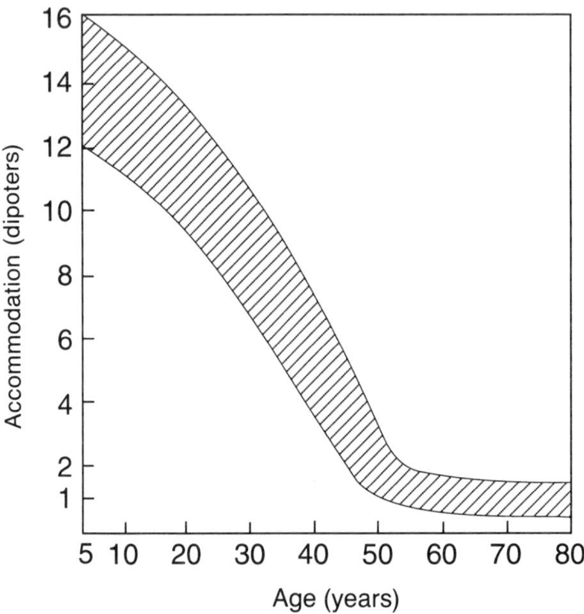

Figure 9.3 Changes in accommodation of the eye with age. The shaded area indicates that there is a large variability between individuals (Handbuch für Beleuchung, 1975)

point of accommodation is then 9 cm, which can be calculated using the equation:

$$f = \frac{1}{D}$$

where f is the focusing distance (metres), and D is the number of diopters of accommodation.

Likewise, if the far point for a 50-year-old with 2 diopters of accommodation is at infinity, then the near point is 50 cm (Figure 9.4). But assuming that the same 50-year-old has 3 diopters of uncorrected short-sightedness (myopia), then the far point (without glasses) is 33 cm and the near point is 20 cm. A person who is myopic at a young age will typically find that with increasing age the far point moves closer and near point moves further away. For an individual with no refractive errors as a young person, the near point moves further away, while the far point may stay at infinity.

The implication for industrial work is that the different ranges of vision not only affect the visibility of the task but also the work posture. To compensate for their poor vision a myopic (short-sighted) person will move closer, and a hyperopic (far-sighted) individual will move further away. Poor work posture observed in industry is therefore often due to poor vision. If the vision is corrected with eyeglasses, the bad

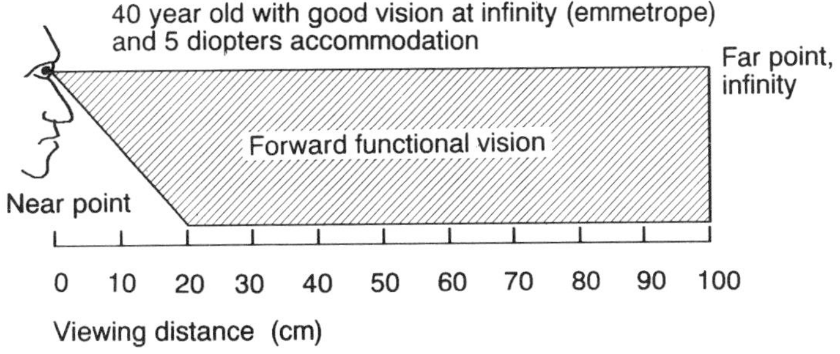

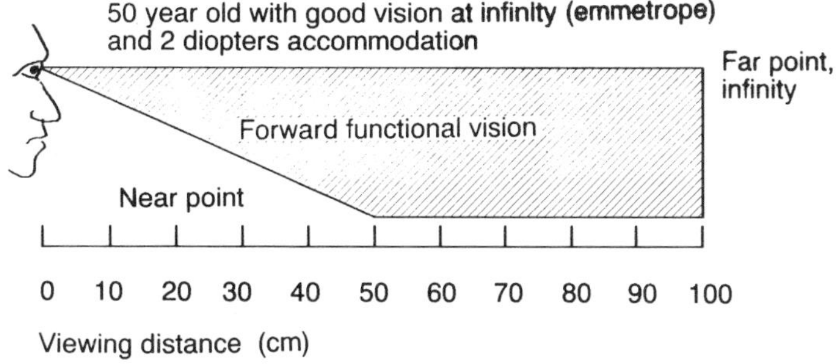

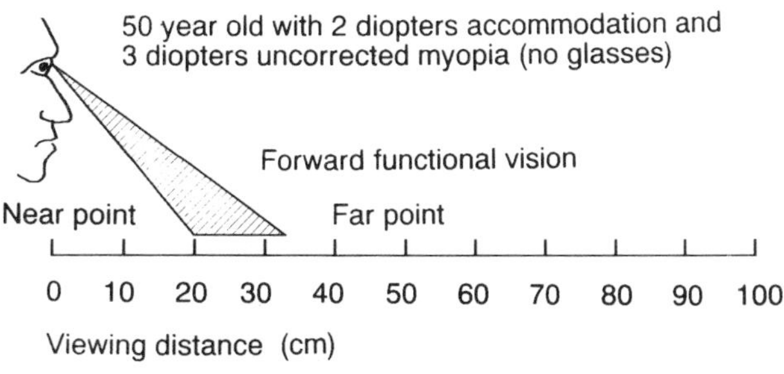

Figure 9.4 Calculation of the near point and far point of 'forward functional vision'. The clear range of vision depends on the range of accommodation (in diopters) of the lens in the eye, and the refractive error

posture may also correct itself automatically. Workers are often not well informed about what kinds of visual corrections are feasible. To help in advising workers some companies hire optometrists who measure the exact viewing distances from the eyes to the various task elements. Eyeglasses which are tailored to the conditions at work can then be prescribed.

The limited range of clear vision makes it necessary that items in a workplace are put at a distance where they can be seen clearly. In the same way that forward functional reach limits the physical organization of a workspace, so does *forward functional vision*.

The second most important affect of age is the *clouding of vision*. In the vitreous humor, between the lens and the retina, there are particles and impurities. With age these impurities increase in size. They impair clear vision, because they scatter incoming light over the retina. Older persons are therefore particularly sensitive to glare sources or stray illumination, which add a veiling luminance over the retina (Wright and Rea, 1984). As a result, the contrast on the retina decreases. For older persons it is therefore important to minimize stray illumination and glare that is not part of the task (Figure 9.5).

Direct glare comes from light sources, such as overhead luminaries, that are shining directly into the operator's eyes. The reflected or indirect glare is from light that is reflected in the workplace from glass or plastic covers, shiny metal, or key caps on a keyboard. One way of solving the problems of both direct and reflected glare is to use task illumination. This involves directing lamps with a restricted light cone towards the visual task. Some examples of task lights are shown in Figure 9.6 (Carlsson, 1979).

9.6 Use of Indirect (Reflected) Lighting

Many office architects and interior designers prefer to use indirect (reflected) lighting because it creates a pleasant environment (Carlsson, 1979). In this case, about 65% of the illumination is directed upwards to the ceiling and then reflected from the ceiling back to the workplace (Figure 9.7). The use of indirect lighting minimizes both direct glare and indirect glare. It minimizes direct glare, because the light is directed towards the ceiling rather than the operator's eyes, and it minimizes reflected glare because the light reflected from the ceiling is non-directional, and will generate so-called 'diffuse reflection'.

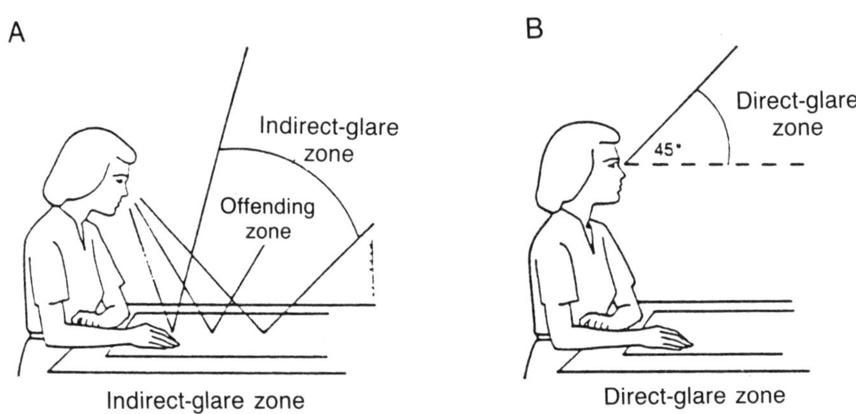

Figure 9.5 Indirect glare (A) arises from reflected light, whilst direct glare (B) arise directly from the light source

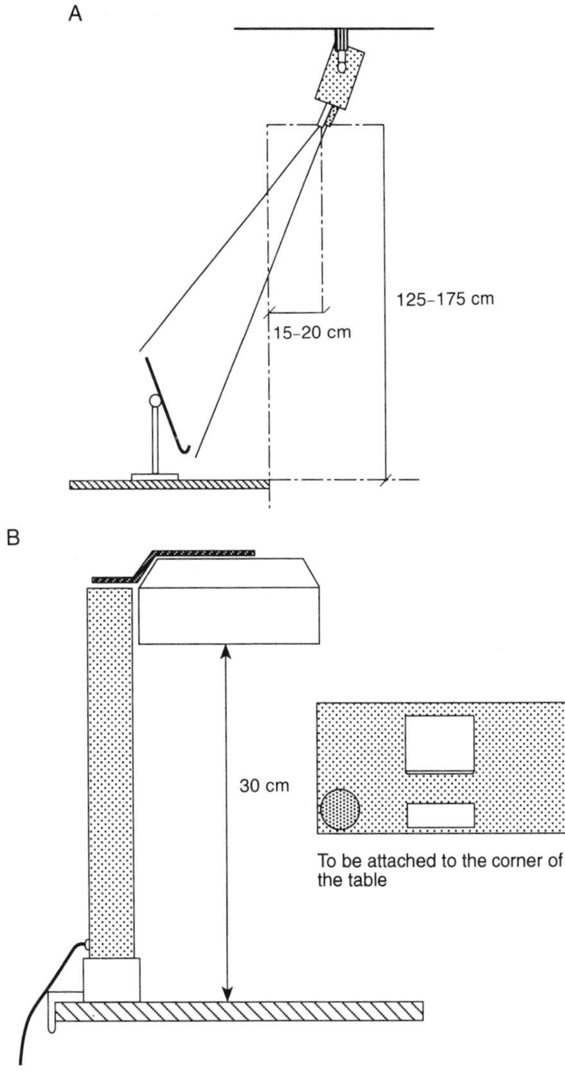

Figure 9.6 Examples of task illumination. (A) An overhead task light with a limited light cone is used to illuminate a source document on a document holder. (B) A table-top lamp can be used on a workbench to provide task illumination that does not generate glare

There is one disadvantage of indirect lighting, namely the loss of light when it is reflected from the ceiling. It is preferable to use a white ceiling, with a high reflectance value. Indirect lighting is mostly suitable for offices and 'clean' manufacturing workplaces where the ceilings do not become soiled. Indirect lighting would probably not be effective for 'dirty' manufacturing processes, since the light sources and light fixtures become covered with dirt and it is necessary to clean luminaires and paint ceilings at regular intervals.

9.7 Cost Efficiency of Illumination

Konz (1992b) has provided convincing arguments that the cost of industrial lighting is minimal. In fact, a generous illumination level costs only about 1% of the worker's salary (in the USA). As demonstrated in the case studies in Chapter 2, efficient illumination typically increases quality in manufacturing and manufacturing yield. It is ill advised to cut down on the illumination to save a few pennies.

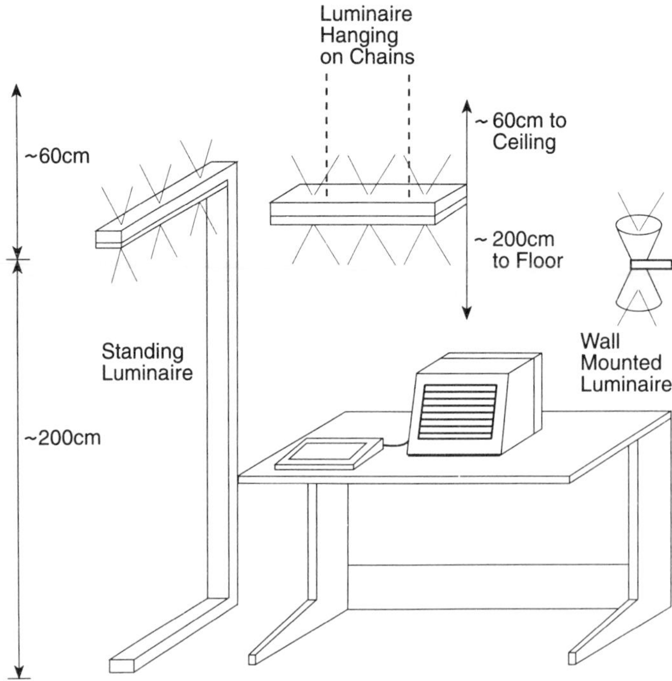

Figure 9.7 Three different types of indirect luminaires. The use of indirect light creates a pleasant atmosphere. About 60–65% of the light is directed upwards and reflected diffusely downwards

The efficiency of a light source is measured in lumens per watt (l m W^{-1}). As illustrated in Table 9.3, some light sources are very efficient whereas others are less efficient. But there is an important trade-off, namely the colour rendering of the light. The colour rendering index (CRI) is a measure of how colours appear under a light source as compared with daylight (or a standard light). A perfect CRI score

Table 9.3 Efficiency of light sources and their colour rendering index (CRI) (adapted from Wotton, 1986)

Type	Efficiency (lm W^{-1})	CRI*	Comments
Incandescent	17–23	92	The least effective but most commonly used light source
Fluorescent	50–80	52–89	Efficiency and colour rendering vary considerably with type of lamp
Coolwhite Deluxe		89	
Warmwhite Deluxe		73	
Mercury	50–55	45	Very short lamp life
Metal halide	89–90	65	Adequate colour rendering
High pressure sodium	85–125	26	Very efficient, but poor colour rendering
Low pressure sodium	100–180	20	Most efficient, but extremely poor colour rendering. Used for roads

*The maximum value of the CRI is 100.

is 100. The main concern is that the colour of the light may distort perception. Low pressure sodium light, which is intensely yellow, makes faces look grey and should not be used indoors. It is mainly used for outdoor lighting, but even in this situation it is difficult to, say, find a car in the car-park because all colours look similar. Measures of light source efficiency and colour rendering are also presented in Table 9.3.

The incandescent light produces the best colour rendering, so that faces look natural, but the efficiency is only 17–23 1 m W^{-1}, which makes this light expensive to use. Fluorescent lights have fair to good colour rendering. The best colour rendering is obtained with the Coolwhite Deluxe source, which has more red colours in the spectrum and looks more natural. The light efficiency varies quite a lot (50–80 1 m W^{-1}). The other light sources (mercury, metal halide, and high pressure sodium) have fairly poor colour rendering and should not be used in manufacturing plants or offices. They are more appropriate in environments where there are few people, e.g. in warehouses, shipping and receiving, and outdoors (Wotton, 1986; Boyce, 1988).

9.8 Special Purpose Lighting for Inspection and Quality Control

Special types of illumination can be used to detect faults in manufacturing. For example, to make surface scratches on glass or plastic visible, it is common to use edge lighting that is directed from the side. There are many other special types of light, including polarized lights, cross-polarization, spotlights, convergent lights, and transillumination (Faulkner and Murphy, 1973). The information given in Table 9.4 is adapted from Eastman Kodak (1983), where more complete information is given. The second column in the table describes special purpose lights or other aides, and the last column describes how the techniques work.

Table 9.4 Special-purpose lighting for inspection tasks (adapted from Eastman Kodak Co. (1983), with permission)

Desired enhancement in inspection task	Special-purpose lighting or other aids	Technique
Enhance surface scratches	Edge lighting can be used for a glass or plastic plate at least 1.5 mm thick	Internal reflection of light in a transparent product; use of a high-intensity fluorescent or tubular quartz lamp
	Spotlight	Assumes linear scratches of known direction; provide adjustability so that they can be aligned to one side of the scratch direction; use louvres to reduce glare for the inspector
	Dark-field illumination (e.g. microscopes)	Light is reflected off or projected through the product and focused to a point just beside the eye; scratches diffract light to one side
Enhance surface projections or indentations	Surface grazing or shadowing	Collimated light source with an oval beam
	Moiré patterns (to accentuate surface curvatures)	Project a bright collimated beam through parallel lines a short distance away from the surface; looking for interference patterns (Stengel, 1979); either a flat surface or a known contour is needed

Table 9.4—Continued

Desired enhancement in inspection task	Special-purpose lighting or other aids	Technique
	Spotlight	Adjust angle to optimize visualization of these defects
	Polarized light	Reduces subsurface reflections when the transmission axis is parallel to the product surface
	Brightness patterns	Reflection of a high-contrast symmetrical image on the surface of a specular product; pattern detail should be adjusted to product size, with more detail for a smaller surface
Enhance internal stresses and strains	Cross-polarization	Place two sheets of linear polarizer at 90° to each other, one on each side of the transparent product to be inspected; detect changes in colour or pattern with defects
Enhance thickness changes	Cross-polarization	Use in combination with dichroic materials
	Diffuse reflection	Reduce contrast of brightness patterns by reflecting a white diffuse surface on a flat specular product; produces an iridescent rainbow of colours that will be caused by defects in a thin transparent coating
	Moiré patterns	See 'Enhance surface projections or indentations', above
Enhance non-specular defects in a specular surface, such as a mar on a product	Polarized light	A specular non-metallic surface acts, under certain conditions, like a horizontal polarizer and reflects light; non-specular portions of the surface will depolarize it. Project a horizontally polarized light at an angle of 35° to the horizontal
Enhance opacity changes	Transillumination	For transparent products, such as bottles, adjust lights to give uniform lighting to the entire surface; use opalized glass as a diffuser over fluorescent tubes for sheet inspection: double transmission transillumination can also be used
Enhance colour changes, as in colour matching in the textile industry	Spectrum-balanced lights	Choose lighting type to match the spectrum of lighting conditions expected when the product is used; use 3000 K lights if the product is used indoors, 7000 K light if it is used outdoors
	Negative filters, as in inspecting layers of colour film for defects	These filters transmit light mainly from the end of the spectrum opposite to that from which the product ordinarily transmits or reflects; this reversal makes the product surface appear dark except for blemishes of a different hue, which are then brighter and more apparent
Enhance fluorescing defects	Black light	Use ultraviolet light to detect cutting oils and other impurities; may be used in clothing industry for pattern marking; fluorescing ink is invisible under white light, but very visible under black light
Enhance hairline breaks in castings	Coat with fluorescing oils	Use of ultraviolet light inspection will detect pools of oil in the cracks

Chapter 10

Design of VDT Workstations

The introduction of visual display terminals (VDTs) to the workplace has given rise to much discussion about potential ergonomics problems including: visual problems, effect of sitting posture, exposure to radiation, and effect of computer usage on job satisfaction. In this chapter we focus on traditional ergonomics problems: sitting work posture, illumination, and screen visibility. Although much of the knowledge on VDT ergonomics has come from research done in office environments (Grandjean and Vigliani, 1980; Grandjean, 1984), VDTs are now commonly used in manufacturing.

Since the introduction of VDTs in the workplace, there have been tremendous developments in technology and the ergonomics of design (Winkel and Oxenburgh, 1993). This is illustrated in Figure 10.1, which was taken in 1982. The figure shows the former US President Jimmy Carter in his home in Plains, Georgia, and illustrates several ergonomics problems, which are explained below.

10.1 Sitting Work Posture

Many of the important design elements of a VDT workstation are defined by Figure 10.2. Some of these design elements have been 'standardized' through national ergonomics standards (Human Factors Society, 1988) as well as international standards (International Standards Organization, ISO Series 9241, 1995). Below we comment on some of the more important design concepts.

10.1.1 Viewing Angle

The centre of the screen should be depressed at a viewing angle of about 25–35° below the horizontal (Hill and Kroemer, 1989). People who sit in an upright posture prefer to look down rather than look up or look straight ahead. In particular, looking up with the head bent back is a common cause of muscle fatigue in the neck.

10.1.2 Thigh Clearance and Low-Profile Keyboards

A person sitting at a desk has limited space for the keyboard and the table top (see Figure 10.2). In 1981, the early German DIN 66234 standard mandated the use of low-profile keyboards and thin table tops (Deutsches Institut für Normung, 1981). The assumption behind this standard was that operators prefer to type with horizontal underarms and 90° elbow angles. If so, the available vertical space between the hands and upper legs can be calculated from anthropometric tables as: sitting elbow height minus thigh clearance (see Table 3.2). For a small 5th percentile female operator this is 7.5 cm – barely enough to fit a 3 cm keyboard and a 3 cm table top.

The German DIN 66234 standard had a pervasive effect since all computer manufacturers complied and manufactured low-profile keyboards (Helander and Rupp, 1984). But there were many protests because the German requirements were perceived as excessive. Ironically enough, a later German investigation proved that the 90°

Figure 10.1 All the ergonomics problems you could think of: (1) Facing a window; (2) keyboard not detachable; (3) no document holder; (4) glare on the screen; (5) chair too low; (6) arm rests interfere with keying; and (7) inadequate leg clearance under the table (courtesy of United Press International)

assumption was indeed excessive. In this investigation muscle activity in the shoulders and the neck of VDT operators was measured using electromyographic activity (EMG) (Zipp *et al.*, 1981). They showed that for elbow angles of 70–90° there was a flat minimum in EMG activity. Thus, it does not seem to matter if the arms are raised to 70° elbow angle. This makes quite a difference in design, since there is no longer a strong argument for a low-profile keyboard. Nonetheless a low profile is a good design feature because it provides greater flexibility in adjusting a VDT workstation to an appropriate height.

10.1.3 Chair Design

In modern office chairs many design features are adjustable. The USA Standard ANSI/HFS 100 mandates adjustability of the seat height over the range 40.6–52 cm (16.0–20.5 in.) (Human Factors Society, 1988).

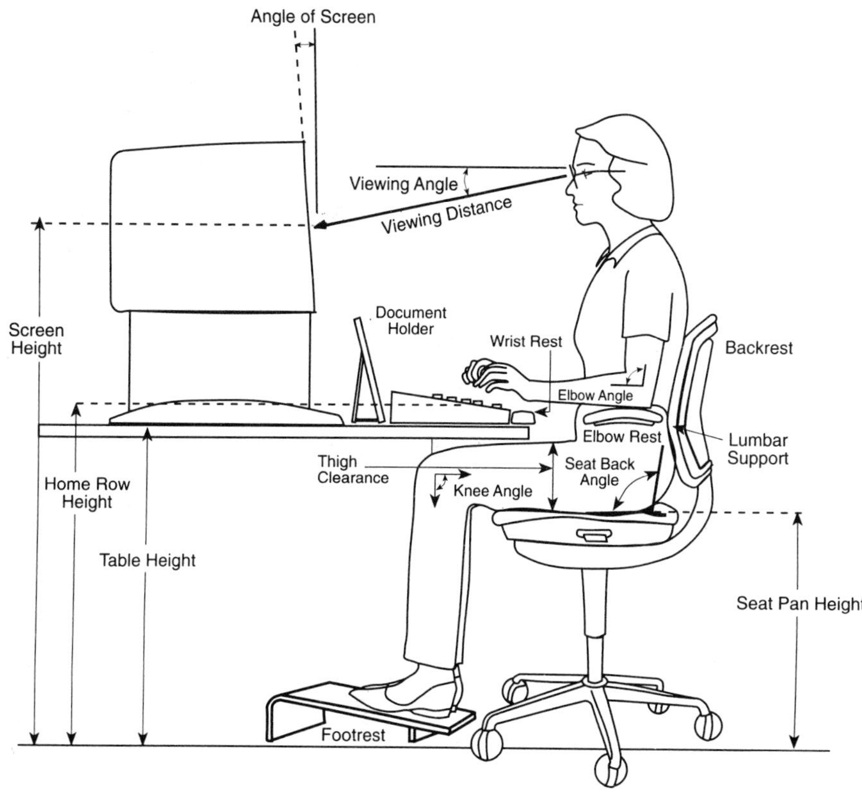

Figure 10.2 Definition of VDT workstation terms

This is the most important adjustability feature and the only one that is mandatory in the ANSI Standard.

The second most important factor is adjustability of the seatback angle. A seatback angle of greater than 110° reduces the pressure on the spine (Michel and Helander, 1994). As a person moves from a straight standing posture to a straight sitting posture, the hipjoint angle goes from 180° to about 90°. The last 30° of movement from about 120° to 90° are absorbed by the pelvis, which rotates forward. This biomechanical change reduces the length of the leverage arm from erector spinae muscles (back muscles) to the spine. As a result the disk pressure is about 30% greater while sitting as compared with standing (Andersson and Örtengren, 1974).

The third most important adjustability factor is the lumbar support. This design feature may have been 'oversold'. Lumbar supports are often not used since chair users do not sit straight and usually do not press their back all the way into the backrest. In fact, many chair users prefer a more relaxed sitting posture (Grandjean, 1986). The lumbar support can become very uncomfortable if it puts pressure on the wrong spot in the back, and lumbar supports must therefore be adjustable (Branton, 1984).

10.1.4 Supports for the Hands, Arms and Feet

A footrest can be helpful for short operators, so that they can support their feet. However, footrests should not be used out of convention. In Figure 10.2 the footrest is unnecessary, since the operator can put their feet on the floor without it.

Arm rests should not interfere with the desk. For a keying task where the operator must pull the chair close to the table, short arm rests (elbow rests) are often preferred over long arm rests.

Wrist rests are optional. Because typing habits are different, some operators prefer wrist rests and some do not. Soft wrist rests (rather than hard) are supposed to put less pressure on the wrist and reduce the risk of developing carpal tunnel syndrome. However, research has not been able to prove there are any significant benefits of wrist rests, and whether soft wrist rests really make a difference. Footrests, armrests and wristrests are inexpensive, and it is good practice to make them available to operators who ask for them (Sauter *et al.*, 1985).

10.1.5 Viewing Distance

Some researchers claim that the viewing distances to the screen, to the documents on the document holder, and to the keyboard should be identical so that it is not necessary to refocus the eyes. Refocusing takes time and is unproductive (Cakir *et al.*, 1980). In addition, for older operators with presbyopia (inflexible lens in the eye) the uniform distance is helpful since it is easier to focus. Another school of researchers claim that it is important to keep exercising the focusing mechanism (accommodation) of the eye. Thereby visual fatigue and 'temporary myopia' can be avoided (National Research Council, 1983). The term 'temporary myopia' implies that the accommodation or focusing of the eyes adjusts to the somewhat closer viewing distance which is imposed by a close working task (Östberg, 1980). Thereby the range of clear vision is moved closer to the eye, and it is difficult to focus on distant objects. This phenomenon is not unique to VDT work. Every close work task may cause temporary myopia, which typically goes away an hour after work. Nonetheless, many individuals notice these effects and are overly concerned. For example, when driving home after work during darkness, temporary myopia combined with dilated eye pupils makes it difficult to read traffic signs. Some individuals may misinterpret this and obtain eye glasses to correct a condition which hardly needs any correction.

10.2 Visual Fatigue

Many VDT operators complain about visual fatigue. However, there are indications that the visual fatigue does not have anything to do with the VDT as such (Helander *et al.*, 1984). Rather, working with VDTs is sometimes a very intense and fatiguing task. For example, a data input operator may input as many as 20 000 characters per hour for 8 hours a day. Typically they look at the source document and glance at the screen only occasionally to check the formatting of the data. After such an intense work day it should not be surprising that operators are fatigued in their entire body, and visual fatigue is just another aspect of general fatigue. Several studies have indeed confirmed that data input operators complain the most about visual fatigue, although this type of work involves comparatively little screen viewing (Helander *et al.*, 1984).

Several investigations have addressed the long-term effect on vision of VDT viewing. Researchers generally agree that there are no adverse effects (Bergqvist, 1986). The eyes do not become more myopic, hyperopic or presbyopic, nor do these conditions develop more rapidly. Most of the changes in eyesight experienced by VDT operators are normal and due to ageing, and they would have happened with any close work.

Nonetheless, VDT viewing is visually more exacting than other visual tasks. Compared with printed characters on a paper, VDT characters are more blurred and there is less luminance contrast between the characters and the screen background. This has been shown to decrease the speed of reading VDT screens as compared with paper (Gould and Grischkowsky, 1984). However, high-resolution screens improve character definition and are easier to read.

Another, more severe, aspect of VDT viewing, is that many individuals lack eyeglasses with appropriate correction (Sauter *et al.*, 1985). This is particularly true for older operators who use bifocal lenses. The lower part of the bifocal lens is typically ground for a viewing distance of about 30 cm and the upper part for a far viewing distance of about 400 cm. The distance from the eyes to the VDT monitor may be around 40–50 cm and the most clear image is obtained if the operator bends his or her head back to read the screen through the lower part of the lens and at the same time moves the head closer. This causes neck strain, and operators often complain about neck pain and shoulder pain (Sauter *et al.*, 1985).

Many companies now supply special glasses known as 'terminal glasses'. An optometrist can measure the viewing distance at the workstation and prescribe lenses which are ground for the exact viewing distance.

10.3 Effect of Radiation

The news media and the popular debate keep bringing up the issue of screen radiation. There is now solid evidence that VDTs do not generate hazardous radiation (National Research Council, 1983; Bergqvist, 1986). In fact, the amount of X radiation, ultraviolet radiation, and infrared radiation are at such low levels that they are difficult to measure, and they are not considered a health risk. Nonetheless, many VDT workers have expressed concern that the exposure to radiation emitted by VDTs might lead to formation of cataracts. Available data indicate that the threshold dose of X radiation that induces cataracts in humans is between 200 and 500 rad for a single exposure and around 1000 rad for exposure spread over a period of several months (National Research Council, 1983). In comparison, a VDT worker exposed to 0.01 mrad/h would absorb less than 1 rad in 40 years of work at a VDT. Likewise, the level of ionizing radiation generally believed to increase significantly the risk of birth defects is more than 1 rad for acute exposure. In contrast a worker exposed to VDT work would absorb 14 mrad over a period of months. There are indeed many other items in our daily life which generate more X radiation than VDTs, including brick walls and self-illuminating dials on wristwatches.

One remaining concern is the effect of electromagnetic radiation on VDT operators. One problem in assessing the effects has been the lack of basic research to prove the effect of electromagnetic radiation on laboratory animals and organisms in general. However, some recent research indicates that railway engineers who are exposed to a large amount of electromagnetic radiation from overhead power lines as well as train engines may have an increased risk of leukaemia and pituitory cancer (Floderus *et al.*, 1993). It is unlikely that VDT operators, who experience much lower radiation levels, would be at any risk.

10.4 Reducing Reflections and Glare on VDT Screens

VDT screens present special problems, since glare and reflections on the screen may make the text difficult to read. Several national and international ergonomic standards have observed this problem and have proposed guidelines for designing workplaces so as to maximize the visibility of the screen (Human Factors Society, 1988).

The ideal working environment for a VDT screen is a pitch-black room. The absence of all other light will enhance the screen contrast and make characters very visible. However, this is not very practical since there are other important tasks which do require ambient illumination, including the communication with co-workers.

As the illumination level in a workplace increases, so will the amount of glare on the screen. There are several ways of reducing reflections on VDT screen (Table 10.1).

1. *Cover windows* completely or partially by using draperies, vertical louvres, horinzontal louvres, or a grey film. Vertical louvres are typically preferred over horizontal louvres, because they can be positioned to block the sun, yet permit most workers to look outside. Horizontal louvres often totally shield off the outside view. Windows can also be covered with a neutral density film (usually a grey sheet of plastic) to reduce the transmittance of light from the outside.
2. *Place light fixtures strategically*. Figure 10.3 provides a side view and a view from behind of an operator at a VDT workstation. In the figure we assume that the light fixtures have a restricted light angle α of about 100°. This may be typical for 'egg crate' luminaires. The operator in Figure 10.3(A) sits at the borderline location of luminaires A and C where there is no direct glare from luminaire C and no reflected glare from luminaire A. Figure 10.3(B) illustrates that luminaire B2, which is closer than the other luminaires, will cause more veiling reflections and wash out more contrast on the screen. Locations B1 and B3 are better. In summarizing the points made in Figure 10.3, luminaires should be placed to the side of operators and not at the front or the back, where they cause more direct glare, indirect glare, and veiling luminance.
3. *Use directional lighting*. The examples in Figure 10.3 illustrate the use of directional lighting or task lights.
4. *Move the workstation*. A VDT operator should not face a bright window, since the large contrast between the dark screen and the bright window may cause discomfort due to glare. Nor should an

Table 10.1 Eight ways of reducing screen reflections (adapted from Helander, 1986)

Location	Measure
At the source	1. Cover windows partially 2. Place light fixtures strategically 3. Use directional lighting
At the workstation	4. Move the workstation 5. Tilt the screen 6. Use screen filters or coatings 7. Use reversed video
Between the source and workstation	8. Hang or erect partitions

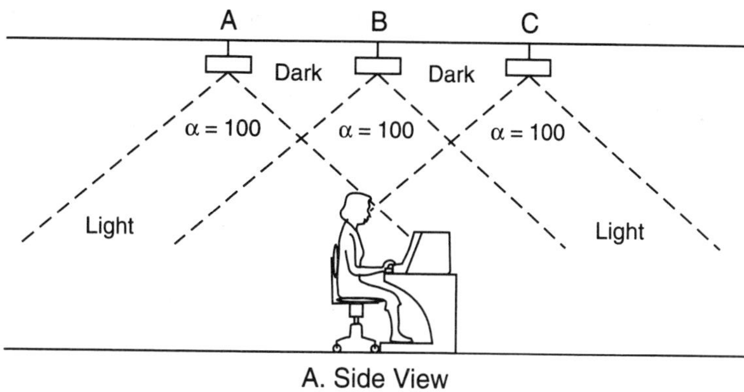

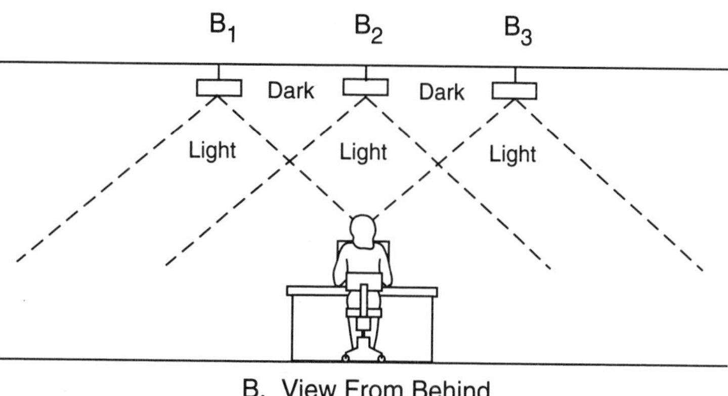

Figure 10.3 From these two figures one can conclude that the ideal location of luminaires is to the side of the operator

operator work with his or her back against a window, as screen reflections from the window are inevitable. Rather, the screen should be positioned at 90° to the window. Workstations can also be moved from a bright area to a darker area in an office. This will reduce veiling screen luminance and wash-out of contrast.

5. *Tilt the screen*. The tilting mechanism, which is mandatory in VDT standards, makes it possible to angle the screen so as to avoid reflections from overhead luminaires and other light sources. Just as with a tilted mirror, one can decide what to look at and what not to look at!
6. *Screen filters or coatings*. Table 10.2 gives an overview of different screen treatments and filters. The filters, such as the neutral density (grey) filter, colour filter, and polarized filter, enhance the contrast between characters and background. The incoming illumination is filtered twice: the first time on its way to the screen, and the second time after being reflected by the screen. However, the character luminance is filtered only once (Figure 10.4).

   Application of an etching or frosting to the screen surface reduces specular (mirror-like) reflections. It is no longer possible to see clearly any reflections of one's clothes or face or overhead luminaires, since the reflections become fuzzy. Unfortunately, the screen characters also become a bit fuzzy.

Table 10.2 Overview of antireflection screen treatments (adapted from Helander, 1986)

Treatment	Advantage	Disadvantage
Etching or frosting	Reduces specular reflections	Increases character fuzziness
Quarter-wave or thin-film coating	Reduces specular reflections	Easily scratched or marred
Neutral filter	Increases contrast*	Decreases character luminance
Coloured filter	Increases contrast*	Decreases character luminance slightly
Circular polarizer	Increases contrast*	Decreases character luminance; highly reflective
Micro-mesh or micro-louvre	Increases contrast	Limited angle of view; collects dirt

*Can also reduce specular reflections if treated with etching, frosting, a quarter-wave, or a thin-film.

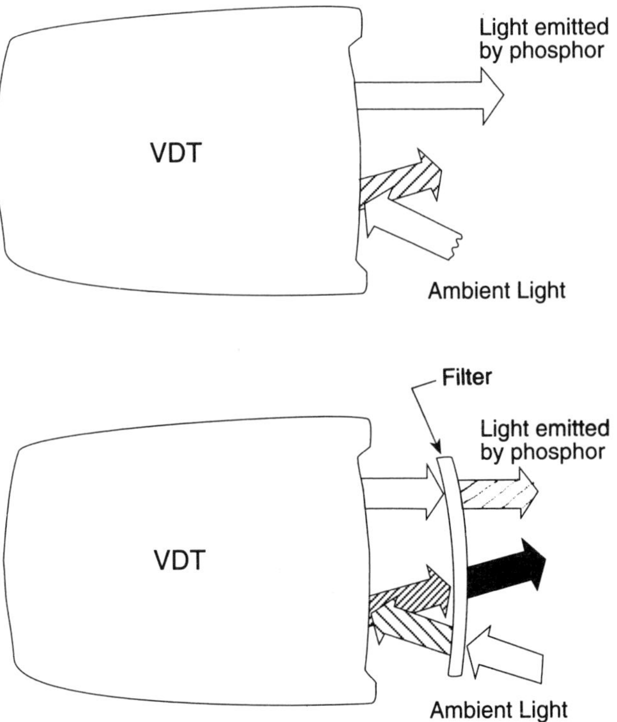

Figure 10.4 A filter increases the contrast between the characters and the background

The quarter-wavelength or thin-film coating is the same type of coating that is applied to camera lenses. Because it is a quarter-wavelength thick, the difference in the distance travelled between the first and the second reflection is a half wavelength and the reflected light is therefore extinguished. This is the most effective antireflection treatment. It is unfortunately a bit difficult to maintain, since the coating is very sensitive to scratches, and thumb prints.

The micro-mesh and micro-louvre filters increase contrast, because there is less ambient (veiling) illumination falling on the screen. The main disadvantages are that the filter limits the viewing angle of the screen, and the fact that the filter collects dust due to build-up of static electricity. It must therefore be cleaned at regular intervals, something that many VDT operators do not think about.

7. *Use reversed video*. Reversed video, meaning black characters on a light background, has become popular because it looks like regular print on a sheet of paper. Reversed video might reduce annoyance since reflections cannot be seen against the white screen background. However, the reflections are still visible against the dark characters, and readability does not improve.
8. *Hang or erect partitions*. By hanging partitions from the ceiling or standing them on the floor, it is possible to block off illumination from light sources in an open-plan office or plant.

10.4.1 Example: Calculating the Effect of a Neutral Density Filter on the Display Contrast Ratio

The purpose of this exercise is to understand illuminance, luminance and contrast ratio, how they are related, and to gain an appreciation of how screen filters work. We use Figure 10.4 as a basis for calculating how a neutral density filter can improve the contrast ratio between the luminance of the characters and the screen background. The ambient light is reflected by the phosphor coating on the back of the screen (this is why the screen surface becomes lighter). We assume that the screen phosphor has a reflectance of 60%, and the neutral density filter has a 50% transmittance.

(1) Calculate the contrast ratio without the filter. If the incident ambient illumination is 200 lux the reflected screen luminance can be calculated:

$$\text{Reflected screen luminance: } \frac{200 \times 0.6}{\pi} = 38.20 \text{ cd m}^{-2}$$

Assuming a phosphor luminance of 300 cd m^{-2}, the character luminance is obtained by adding the two contributions:

Character luminance = 338.2 cd m^{-2}

The contrast ratio (C_r) is then:

$$C_r = \frac{338.2}{38.2} = 8.9$$

(2) Calculate the contrast ratio with the 50% filter.

$$\text{Reflected screen luminance: } \frac{200 \times 0.5 \times 0.6 \times 0.5}{\pi} = 9.6 \text{ cd m}^{-2}$$

The contrast ratio is then:

$$C_r = \frac{300 \times 0.5 + 9.6}{9.6} = \frac{159.6}{9.6} = 16.7$$

Thus the contrast ratio is almost double that which enhances visibility. The only disadvantage is that the character luminance has decreased from 338.2 to 159.9 cd m^{-2}. In our case this reduction is not critical, since 159.9 cd m^{-2} gives very good visibility.

Chapter 11

Design of Controls

As with anthropometry, a great deal of research on control design has been sponsored by the US Department of Defense. The main purpose has been to develop design principles that can be used to standardize control design (e.g. Military Standard 1472D, US Department of Defense, 1989). In this chapter we draw from these design principles as they can be applied to manufacturing. Particularly useful are the principles of 'coding of controls', 'control movement stereotypes', and the 'control–display relationship'.

In the manufacturing scenario, operators handle a variety of objects – not only controls. There are hand tools and parts to be assembled. The design principles listed here can be extended to parts and hand tools also – in fact they apply to anything that is handled purposefully in manufacturing. For example, the principles of 'coding of controls' can be extended to 'coding of parts' and 'coding of hand tools'.

We first provide some principles for choosing controls and then examine computer input devices.

11.1 Appropriateness of Manual Controls for the Task

Manual controls should be selected such that they are appropriate to the task and intuitive to use. Some controls can make a task easy to perform, whereas others make a task more difficult. One way of analysing control requirements is shown in Table 11.1, where controls are classified by the number of control settings and by the force required to manipulate the control. For example, if a control does not require much force and there are only two discrete settings, such as an on–off control, the recommended types are: toggle switch, push-button, or key lock. If there are several control settings, a rotary selector would be a good device.

If greater actuation force is necessary one must select a control where it is easy to apply more force. Finger-actuated controls would not do, but rather controls such as hand push-buttons, foot pedals, levers or cranks should be used. Note that in Table 11.1 some of the controls are for continuous settings, and in this case joysticks or handwheels could be appropriate.

Over the years controls for certain environments have become fairly standardized. This is, for example, the case for automobile controls. We have become so accustomed to steering wheels and foot pedals, that it would be difficult to imagine any other arrangements. Some examples of standardized controls are:

- Steering wheel for steering.
- Joystick for three-dimensional vehicle steering.
- Foot pedals for break and acceleration.
- Manual lever for aircraft throttle.
- Lever control for gear shift.

Table 11.1 The choice of control depends on the force and the number of control settings (adapted from Chapanis and Kinkade, 1972)

Forces and settings	Control
Small actuation force	
2 discrete settings	Key lock, push-button, toggle switch
3 discrete settings	Rotary selector, toggle switch
4–24 discrete settings	Rotary selector switch
Small range of continuous settings	Knob, joystick lever
Large range of continuous settings	Crank, rotary knob
Large actuation force	
2 discrete settings	Hand push-button, foot pedal
3–24 discrete settings	Detent lever, rotary selector switch
Small range of continuous settings	Handwheel, joystick lever
Large range of continuous settings	Crank, handwheel

- Rotary control for continuous panel control functions.
- Rotary knobs or T-handles for manual valve control.

We have no argument with the use of these controls as they have become *de facto* standards; an operator would be confused and annoyed to find other types of control. But many of these industry standards were created at the beginning of the century, and were put in place without any experimental evidence of what type of control is best.

Today we understand better how to evaluate controls, and if we were to select a control for steering an automobile, there would probably be extensive research. One example of such research was for the design of push-buttons for telephones. Bell Laboratories of AT&T actually investigated nine different alternatives for the layout of push-buttons. Figure 11.1 shows two of the alternatives. The 'telephone layout' was chosen over the 'calculator layout', although the latter was already a *de facto* standard in offices. The main reason was that experiments showed that users make fewer errors in dialing with the 'telephone layout' than with the 'calculator layout'. Since the dialing of wrong telephone numbers is very costly to the telephone system,

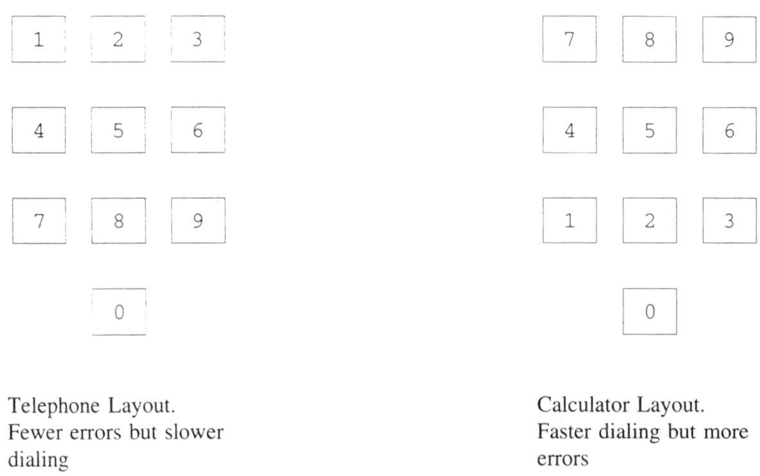

Figure 11.1 Two alternative layouts of a telephone control set

the 'telephone layout' won out, although it is actually a little slower for dialing.

As a result we have now the confusing situation that there are two types of numeric keyboard in an office; confusing but considering the cost trade-off, probably the best overall design.

11.2 Computer Input Devices

With the emergence of computers, control functions have become more abstract. Systems that used to be controlled manually can now often be controlled by a computer. This scenario opens up new design options for controls. For example, it may be possible to use a joystick to select an item on a menu, or type command words, or mouse-click on certain areas of the screen. The control action and the system response can then be represented graphically on a computer screen.

There has been much research into computer input devices, most of it sponsored by computer manufacturers. There are six well-established input devices: mouse, track-ball, joystick, touch screen, light pen, and graphics tablet. Some of the advantages and disadvantages of these devices are summarized in Table 11.2 (Greenstein and Arnaut, 1988).

Different input devices have different advantages and disadvantages, and we briefly comment on these. Pointing devices such as touch screens and light pens provide for excellent eye–hand coordination. Pointing is a very direct way of expressing preference. A child will point at something and say 'I want that'. This is such basic behaviour that training is not necessary. Touch screens and light pens are therefore the most 'direct' devices (Whitefield, 1986). There are, however, two concomitant disadvantages. The pointing finger will partly obscure the view of the display, and the input resolution is poor. For the touch screen the resolution is the width of the finger, and for the light pen it is the width of the pen. Users can therefore specify only a limited number of options on the screen. Touch screens are particularly appropriate for harsh environments such as manufacturing.

Table 11.2 Advantages and disadvantages of standard pointing devices

	Mouse	Track-ball	Joystick	Touch screen	Light pen	Graphics tablet
Eye–hand coordination	0	0	0	+	+	0
Training requirements	0	0	0	+	0	0
Unobstructed view of display	+	+	+	–	–	+
Input resolution capability	+	+	+	–	–	–
Flexibility of placement within workplace	0	+	+	–	–	0
Space requirements	–	+	+	+	+	–
Comfort in extended use	0	+	+	–	–	0
Capability to emulate other devices	0	0	0	0	0	+
Suitability for pointing	+	+	–	+	+	+
Pointing with confirmation	+	0	0	–	0	0
Drawing	0	–	–	–	–	+
Continuous tracking	0	0	+	–	–	0

+, Advantage; 0, Neutral; –, Disadvantage.

There are no moving parts, they are sturdy and robust and can survive in a contaminated environment.

The mouse, the track-ball and the joystick have the best input resolution. This is because the input resolution can be programmed by changing the 'gear ratio' between the device movement and the cursor movement.

For some devices there is flexibility in where they can be physically located. Track-balls and joysticks are in this respect superior; they are small and easy to move. One disadvantage with the mouse and the graphics tablet is that they occupy prime table space, that is also used for note taking and other tasks. When it comes to operator comfort, the track-ball and joystick are considered superior, particularly to touch screens and light pens: sitting with an outstretched arm causes muscle fatigue.

Graphics tablets are primarily used for drawing. In addition a graphics tablet can be programmed with special functions or subroutines (like a CAD tablet). For example, a manufacturing process can be depicted on the tablet, and machine commands can be programmed on the tablet. Thereby, a graphics tablet (for input), and a corresponding display (for feedback), can emulate a manufacturing process. It is a very functional and naturalistic input device, since there can be a one-to-one correspondence between the input, display feedback and what happens in real life.

There are many different input tasks including pointing, pointing with confirmation (double click), drawing, and tracking. The main advantage of a mouse is for pointing with confirmation; track-balls are less appropriate for this. Joysticks are superior for military tracking tasks (Parrish *et al.*, 1982). Touch screens and light pens are primarily good for pointing, because they are very intuitive.

Although it seems fair to characterize input devices as discussed above, new inventions are continually being advertised in computer magazines. Before one can make any judgement about these novel devices, it is necessary to test them. This would imply experiments with human test subjects to perform tasks such as pointing, drawing and tracking. The best device would be the one that requires the least time and produces the least number of errors in performing the task.

11.3 Control Movements Stereotypes

People have clear expectations of what to do with controls. In the USA, to turn on a light the expectation is to turn a switch up. For a person raised in Europe, there is the opposite expectation, since the switches go down. The point is that control movement stereotypes are trained expectations, and many have been learned since childhood. Some of the most common stereotypes are shown in Table 11.3. For example, to turn something on there are expectations of a control movement either up (in the USA), to the right, forward, or clockwise (Van Cott and Kinkade, 1972).

To raise an element vertically, such as an overhead crane in a plant, we would expect to move a control vertically upwards, as can be done with a control that extends horizontally. This is a clear stereotype, since there is a one-to-one correspondence between the movements of the control and the controlled element. If the lever extends vertically, however, to raise the crane the best stereotype would be to pull the lever back – just as a joystick in an aeroplane is pulled back to raise the aeroplane. But this is a less clear stereotype. The control

Table 11.3 Control movement stereotypes - common expectations for control activation

Controlled element	Human control action
On	Up, right, forward, clockwise
Off	Down, left backward
Right	Clockwise, right
Raise	Up, back
Lower	Down, forward
Retract	Up, backward, pull
Extend	Down, forward, push
Increase	Forward, up, right, clockwise
Decrease	Backward, down, left, counterclockwise
Open valve	Counterclockwise
Close valve	Clockwise

movement is horizontal, but the controlled element moves vertically. Many individuals would make a mistake by pushing the control forwards (unless they have been thoroughly trained). Thus our first option for a crane control would be a horizontal control lever.

The stereotype to open a valve is always to turn it counterclockwise (unscrew) and to close the valve the control is turned clockwise. It would seem that manufacturers of bathroom taps could learn from this principle. There is now a proliferation of different designs, and unaccumstomed users (e.g., hotel guests) cannot understand how to operate the controls. Many accidents happen in bathrooms when old people, in particular, slip and fall because they are startled by the water temperature. Standardization of tap design would be very helpful.

11.3.1 Example: Controls for an Overhead Crane in Manufacturing

Controls for overhead cranes in manufacturing often violate control movement stereotypes. The author was recently involved as an ergonomics expert in litigation. In this case a worker had been injured trying to catch the hook of an overhead crane when the hook was lowered. He was not successful in catching the hook, lost his balance and fell about 3 m to the floor, severely injuring his hip. The hoist was being lowered by a fellow worker using a control box which was strapped to his stomach (Figure 11.2B). On the control box there were four identical lever controls with a neutral detent position. The main hoist and the auxiliary hoist conformed with common stereotypes (push to lower). However, movement of the hoist along the bridge and the trolley was controlled by moving the levers in the same direction, which is confusing. Ideally, these controls should have been operated by a joystick which could move the hoist in the x and y directions simultaneously.

This was not the only problem. There was a second set of controls in a crane cab which was located under the ceiling of the manufacturing facility. The four controls in the cab were obviously confusing, since someone had pasted labels next to the control levers to indicate how they were supposed to be moved. For three of the controls (the auxiliary hoist, the main hoist and the trolley) there was actually compatibility with the direction of movement, but for the bridge control there was not. A forward movement of the lever made the bridge move to the right and a backward movement made the bridge move to the left (Figure 11.2C).

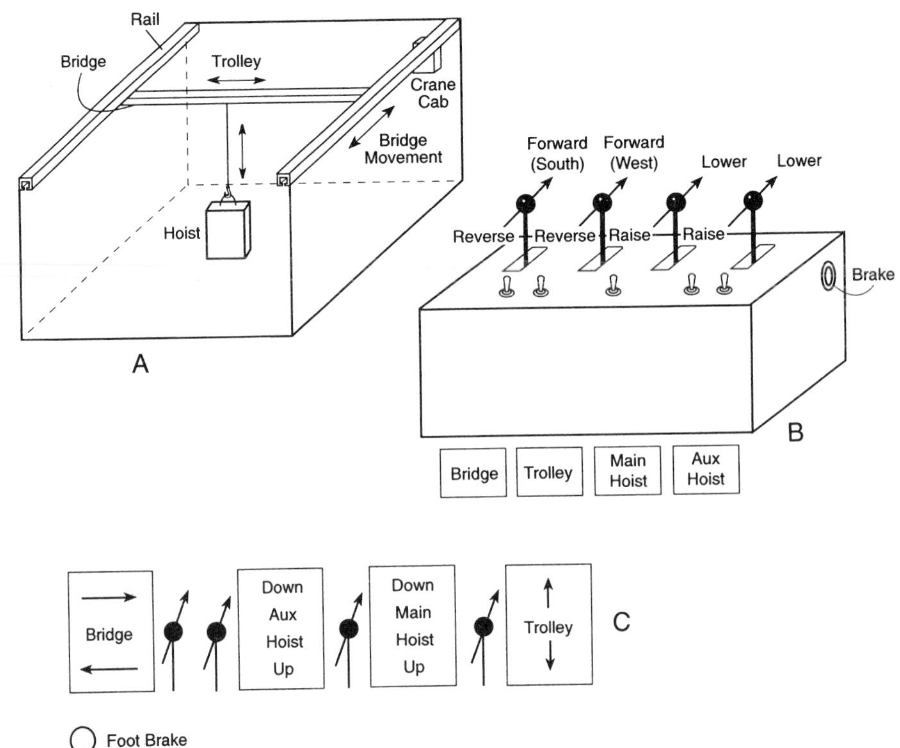

Figure 11.2 (A) The layout of the manufacturing area – note the crane cab in the top right-hand corner. (B) The control box, carried with a strap on the stomach. (C) Layout of the control arrangement in the cab – note the labels which were taped next to the controls

A third problem was that the layout and the ordering of the controls in the crane cab were *different* from the hand-held control box. An operator who was used to the control arrangement in the cab would have problems using the control box, and vice versa, and there would be an increased likelihood of errors in activating the control. In the human factors literature this is referred to as *negative transfer of training* (Wickens, 1992; Patrick, 1992). Operators are likely to revert to the type of behaviour they learned first, especially when under stress. One can therefore expect many more errors in an emergency situation, and one error is likely to lead to another. It is therefore very important to use control movement stereotypes and to analyse the compatibility between control movements and the controlled element.

There are also advantages for productivity. A good control arrangement will always save time in performing a task. The operator will of course learn the task anyway, but even after several years of practice there will be a remaining advantage in performance time for compatible controls (Fitts and Seeger, 1953).

11.4 Control–Response Compatibility

Chapanis and Lindenbaum (1959) performed a classic study on control–response compatibility. They studied the preferred location of controls for burners on cooker tops. This study was later complemented by Ray and Ray (1979) and their results are represented in Figure 11.3. In the figure the controls are numbered 1, 2, 3, 4 from left to right. There are, however, four alternative layouts of the burners. The question is: which one is the best? To answer this question, Ray and Ray (1979) used 28 female test subjects. There

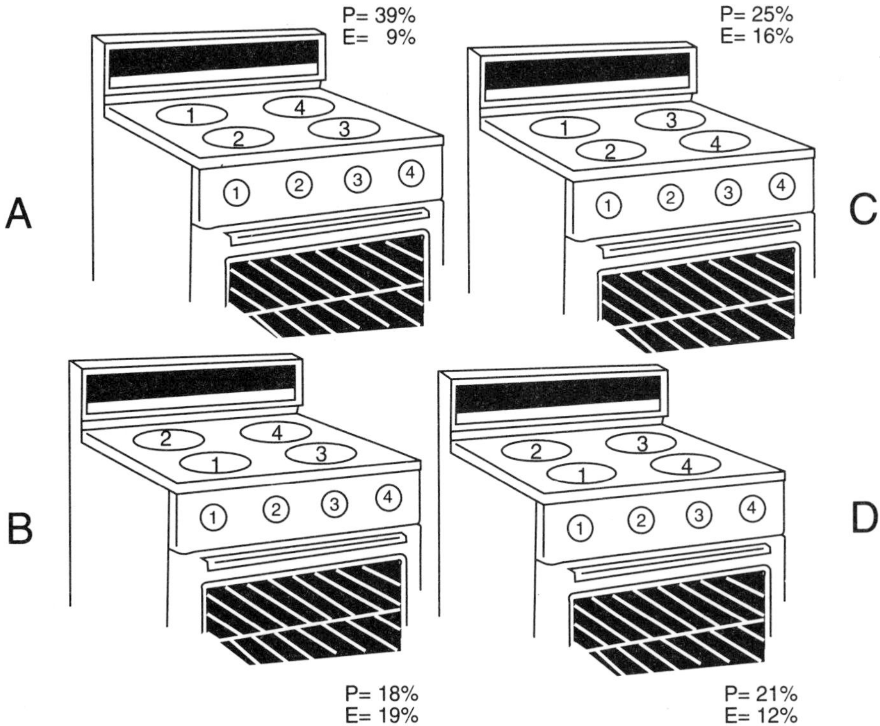

Figure 11.3 The arrangement of controls and burners are used to illustrate the concept of control–response compatibility. Preference (P) and errors in control activation (E) are given as percentages

were 560 trials on each stove, and the preferences (*P*) and errors (*E*) were calculated as percentages for each stove.

The problem with the design of the cooker top is that there is no clear control–response compatibility. Ideally, there should be a one-to-one relationship between the controls and the responses (burners). In our case, the cooker top can be easily redesigned, so that there is control–response compatibility. For example, the rear burners can be offset slightly to the side. The controls can then be lined up one-to-one with the burners, and the association is immediate. Chapanis and Lindenbaum (1959) proved that for this arrangement there was not a single error in control actuation. For a less compatible arrangement such as in Figure 11.3, the user must look a few times to decide. Instead of a simple reaction time there is a double or triple reaction time, which may take 2–3 s rather than 1 s.

The cooker top serves as a familiar example to illustrate control–response compatibility. The point is that manufacturing workstations must also be designed to be compatible. In designing a workstation for manual assembly one must line up part bins so that they are compatible with the assembly process. For example, in the assembly of car brakes (Figure 11.4) there is a one-to-one correspondence between the location of parts in the bins and the corresponding location of parts in the assembly. Such *bin–assembly compatibility* reduces assembly time (Helander and Waris, 1993). Similar principles apply to the design of controls and displays for process control – the controls and displays must be lined up to be compatible.

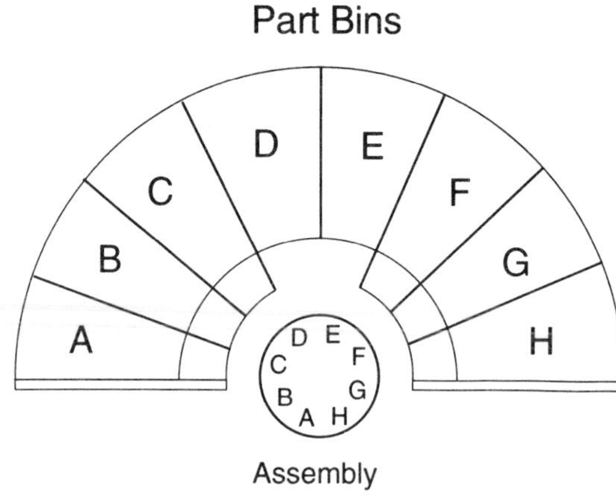

Figure 11.4 Bin-assembly compatibility for the assembly of car brakes, as used at General Motors, Saginaw, Buffalo, NY, USA

11.5 Coding of Controls, Hand Tools, Part Bins and Parts

Controls can be coded by adding features to the controls. This makes them easier to distinguish. There are six common types of control coding (Van Cott and Kinkade, 1972; Sanders and McCormick, 1993):

- Location,
- Colour,
- Size,
- Shape,
- Labelling, and
- Mode of operation.

These principles apply to controls in automobiles and aeroplanes, as well as industrial machines. In manufacturing assembly, 'control coding' can be applied to hand tools, parts bins, and parts - really to anything that is purposefully touched or handled by the operator. We first explain the different types of coding and then give manufacturing examples.

11.5.1 Coding by Location

Coding by location is the most powerful principle. For example, in automobiles the locations of many controls have been well standardized so drivers have clear expectations of where to find certain functions. These expectations build up with increasing experience in driving. Most car drivers can find the location of the ignition blindfolded; the location has been well coded (McGrath, 1976). In manufacturing, standardization of the location of items in a plant would similarly be advantageous. For example, the location of hand tools could be standardized. Thereby operators can rely on location coding to find tools quickly.

11.5.2 Coding by Colour

In colour coding, items are coloured differently depending upon the function and the task. One potential problem with colour coding is that it only works in a well-illuminated environment. Colour coding of controls on underground mining machines, for example, does not make sense.

Colour coding requires a longer reaction time than does location coding, since it is first necessary to reflect on the meaning of the colour before the task can be performed. This typically involves a double reaction time.

Some control colours have stereotypical meaning, and it has become common to make emergency controls red. For example, an emergency stop control for a robot or for a conveyor line must be red. However, different cultures around the world may have different colour stereotypes. Courtney (1986) surveyed a large sample of Hong Kong Chinese to determine the strength of associations between nine concepts and eight colours and then compared these data with a similar study of Americans (Bergum and Bergum, 1981). Results of the two studies are compared in Table 11.4.

There were some substantial differences between the two populations. While 'Cold' is associated with blue among Americans, the predominate colour among the Chinese is white. For the concepts of 'hot', 'Danger' and 'Stop', red was the dominant colour among both populations. However, the percentage values for Americans were much greater than for the Chinese. Courtney pointed out that the reason for the lower percentage values among Chinese is that for them red is primarily associated with 'happiness'. The strength of this association detracts from other possible associations.

Colour coding is also used for displays and for indicator lights. Similar to the coding of traffic lights, green is used to denote a satisfactory condition ('go'). Yellow is used to denote caution or out-of-tolerance conditions, and red is used to denote a hazard or warning condition.

11.5.3 Coding by Size

Control coding by size implies that to distinguish easily between different controls, some control knobs can be made smaller, some medium sized, and some larger. An operator can distinguish, at most, three different sizes of control knobs during conditions of stress - such as fighter pilots in combat (Chapanis and Kinkade, 1972). Therefore, machine controls could be in at least three sizes.

11.5.4 Coding by Shape

Controls can be coded by shape (Figure 11.5). In this case an operator can distinguish up to 12 different shape-coded control knobs under conditions of stress (Woodson and Conover, 1964). The US Air Force has taken advantage of this principle. Manual controls in cockpits typically have different shapes. The best situation is when the control shape resembles the function controlled. In Figure 11.5 the flap control resembles the flap and the landing gear resembles a wheel. But for some (abstract) controls such association is not possible; 'carburettor

Table 11.4 Concepts and most frequently associated colour for two populations: Hong Kong Chinese and Americans

	Chinese		Americans	
Concept	Colour	%	Colour	%
Safe	Green	62.2	Green	61.4
Cold	White	71.5	Blue	96.1
Caution	Yellow	44.8	Yellow	81.1
Go	Green	44.7	Green	99.2
On	Green	22.3	Red	50.4
Off	Black	53.5	Blue	31.5
Hot	Red	31.1	Red	94.5
Danger	Red	64.7	Red	89.8
Stop	Red	48.5	Red	100.0

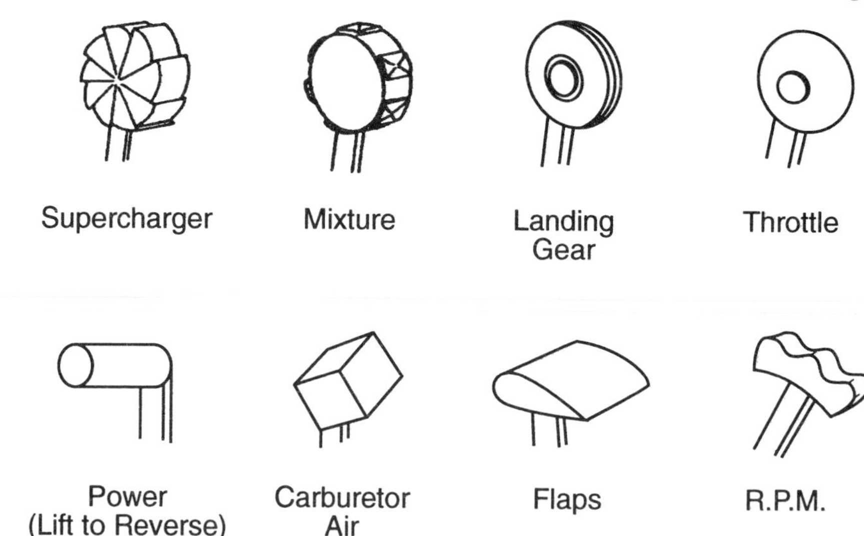

Figure 11.5 Shape-coded controls for aeroplanes

air' is one example and 'revolutions per minute (r.p.m.') is another.

Shape coding of controls can also be used in industrial situations. In fact, sometimes operators add on the shape coding themselves. During investigations following the Three Mile Island nuclear disaster, many intriguing principles of shape coding were discovered to be used by power plant operators. At one nuclear power plant, operators had coded otherwise identical control levers with beer bottles. The bottles were simply put on top of the levers, and this made the controls easier to distinguish.

11.5.5 Coding by Labelling

A written label is used to describe the function. The label can be put above, underneath, or on top of the function. The location of the label does not really matter as long as it is clearly visible and the wording reads from left to right (Chapanis and Kinkade, 1972). Vertical labels take longer to read and should not be used. One problem with labels is that they might not survive in a harsh industrial environment. In particular, printed characters may be soiled or destroyed. Embossed labels are therefore often used in these environments (Loewenthal and Riley, 1980). As with colour coding and shape coding, the use of label coding implies a double reaction time – the label has to be read, and understood before action can be taken.

11.5.6 Coding by Mode of Operation

Finally, controls can be coded by the mode of operation. This implies that controls have a different feel to them or that each control has a unique method of operation. A car driver can distinguish between the accelerator and brake because they have different control resistance, dampening and viscosity. The same principles can be used for controls in industrial settings. The operator can verify that the correct control has been activated and also interrupt a control activation if there is an obvious error.

11.5.7 Coding of Parts and Other Things Touched by the Hand

Many of the coding principles described above can be applied to parts to be assembled and to hand tools. In fact, the principles apply to anything that is touched on purpose. Thus, hand tools can be coded by location, colour, or labelling, and parts can be coded by location, colour, shape or labelling. For example, colour coding can be used as a scheme for organizing a workstation, by applying the same colour to parts bins and hand tools that belong together. Colour coding of parts is nothing new. It has long been used in electronics for marking resistors, transistors and capacitors, and this simplifies electronic assembly. In fact these parts are also shape coded and it would be difficult to confuse a transistor with a capacitor.

11.6 Emergency Controls

The design and location of emergency controls require particular attention, since it is crucial to be able to find them quickly (Etherton, 1986). Emergency situations are stressful and operators are likely to make mistakes. Emergency controls must therefore be particularly well designed to allow fast action without any errors. Some design recommendations are summarized in Table 11.5.

Many types of emergency control are used. In addition to palm-buttons and emergency cords, a 'dead man's' switch may be used. As long as the switch is actively pressed the machinery keeps going. If the pressure is released, the machinery stops. Some types of industrial machinery have an automatic switch or function in case the worker inadvertently comes into the danger area. For example, rotating tyre-building machines have emergency trip cords located above the operator's feet. If the operator is pulled into the machine the feet will catch the trip cord and the machine will stop.

Several additional safety devices that can be used for industrial workstations are described in Chapter 19.

11.6.1 Example: Accidental Activation of Seat Ejection Controls in Aeroplanes

Emergency controls must be placed away from other frequently used controls in a location that is easy to distinguish. This principle has been of great concern in the design of aeroplane cockpits. It used to be the case that fighter pilots were killed by ejecting themselves by mistake when the aeroplane was still on the ground. The pilot would be catapulted 100 m and fall flat to the ground before the parachute had time to open. It also happened that pilots would eject themselves into the ceiling of hangars. The eject control button has now been relocated to a safer place (under the seat between the legs). This location has an additional benefit, since the pilot's arms are kept out of the way during the ejection.

11.7 Organization of Items at a Workstation

All items in a workstation that are handled need to be organized. This includes controls, hand tools, parts to be assembled and part bins. In some cases workers take the initiative to organize a workstation. But one cannot rely on this. It is better if there is a deliberate design

Table 11.5 Recommendations for the design of emergency controls

Position emergency controls away from other frequently used controls, thereby lessening the risk of inadvertant activation.
Make emergency controls easy to reach – put them in a location that is natural for the worker to reach
Make emergency controls large and easy to activate, e.g. use a large rather than a small push-button
Colour emergency controls red

effort, where engineers and workers can collaborate in arranging a workstation.

Predetermined time-and-motion studies (PTMS) such as methods time measurement (MTM), MOST, and WORK FACTOR have primarily been used for predicting and quantifying the time it will take to assemble a product (Konz, 1990). PTMS measurements can then be used to divide a large task into several parts, thus balancing the work between different workers. PTMS is also used as a basis for salary negotiations. However, PTMS could have a much wider usage. It could be used to evaluate the design of a product, and alternatives for organizing a workstation. But this is rarely done - perhaps because there are so many different options for workstation design, it is difficult to understand where to start. Some guiding principles are clearly needed, such as the principles resting on ergonomic knowledge that are described below.

11.8 Principles for the Design of Workstations

1. *Keep the number of items that are touched by the hand to a minimum*. Minimize the number of hand tools, the number of different parts, and the number of controls. The number of parts and the number of necessary tools depend on the product to be manufactured. It is important for product designers to understand the implications of their design in terms of manual labour. Why use five varieties of screw when two are enough? Why not combine parts such as incorporating washers with the screws?
2. *Arrange the items (controls, hand tools and parts) so that the operator can adjust his/her posture frequently*. Sometimes the location of items tie up workers in impossible work postures. There are many examples of industrial machinery which must be operated using a foot control. For example, in using an industrial punch press the operator must hold the work item with both hands and press the foot control to initiate the pressing action. Using just one foot causes one-sided strain that is likely to lead to back problems. It must be possible to move the foot control so that it can be operated with either foot at the worker's convenience.
3. *Consider preferences in hand movements and handedness*. People can move their hands both faster and with much better precision through an arc than they can horizontally or vertically. Imagine that you are drawing a straight line on a piece of paper. It is difficult to get the line straight if it is drawn horizontally or vertically. It is easier to draw if the paper is turned at an angle so that the hand can move outwards from the body, such as in the movement envelopes shown in Figure 11.6. This is because there are only a few active joints in the arm, typically only the elbow joint moves. But for drawing a horizontal or a vertical line there are many more active joints and many muscles that have to interact, which makes the movement more complex.

   Handedness is important in the design of hand tools, particularly those intended for tasks which require skill and dexterity. Assembly tasks do require skill and dexterity, and thus hand tools for left-handed individuals.
4. *Organize items in the workplace*.
   (a) Distinguish between primary and secondary items. Primary items are those that are used most frequently and secondary

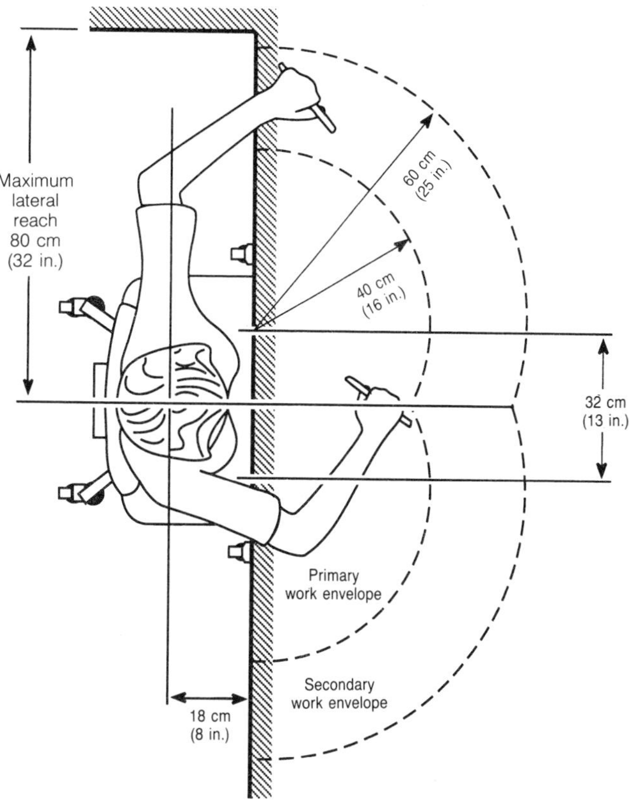

Figure 11.6 Arrangement of a workstation, showing primary and secondary movement envelopes

items are those that are not used as frequently. List all the parts and classify them as primary or secondary items.

(b) Divide the tasks into subtasks, each forming a logical unit. For very short tasks this may not be important, but for more comprehensive tasks it is desirable.

(c) Divide the worktable into several areas, one for each subtask. This may be practical only for comprehensive tasks where there are many items to keep track of. Organizing the items for each subtask is practical and makes it easier for the operator to think of the task.

(d) Identify primary and secondary movement envelopes on the worktable. The functional reach for a 5th percentile female worker is about 40 cm (16 in.) and determines the limit of the primary work envelope (Figure 11.6). Put a primary item in the primary envelope. Secondary items should be put in the secondary envelope, but within a reaching distance of about 60 cm (24 in.).

(e) Locate items such as bins and tools so they can be used sequentially for each subtask. A procedural order helps in organizing the task and facilitates task learning and productivity. A well-organized workstation will save time and is productive. The location of parts, hand tools and controls according to the primary and secondary importance helps in organizing the workstations.

11.9 Recommended Reading

Much information relevant to the design of industrial workstations is available in military design guidelines. Although military tasks are different from industrial tasks, the designer of industrial workstations may consult such resources for ideas. The *Human Factors Design Handbook* (Woodson, 1981) is a comprehensive collection of design guidelines with many illustrations. *Human Engineering Guide to Equipment Design* (Van Cott and Kinkade, 1972) is an inexpensive and useful reference book. An example of the type of information available is given in Figure 11.7.

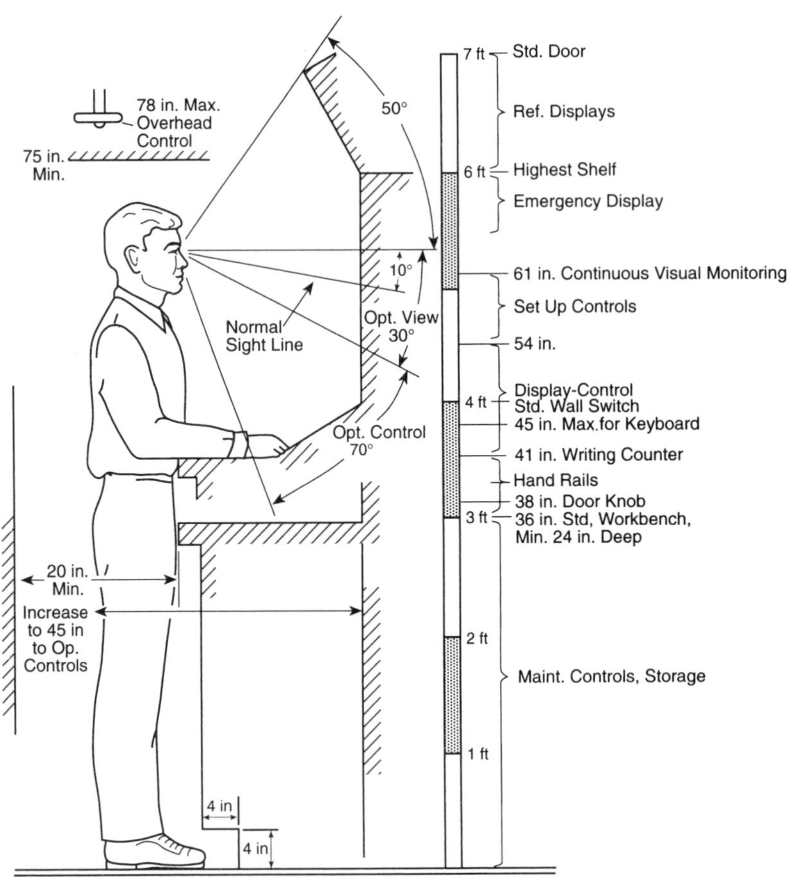

Figure 11.7 Design of a console workstation for a standing operator. Although this workstation was conceived for military applications, it is equally relevant for process control

Chapter 12

Design of Symbols, Labels and Visual Displays

This chapter deals with the problems encountered in designing symbols and text such that they are easy to understand. Symbols and words are often combined in the design of visual displays. In this chapter the term 'visual displays' refers to a wide variety of displays - from posters and signs to computer displays. What these displays have in common is that they carry visual information which must be given a semantic interpretation so that the reader understands what to do. We also discuss the design of safety warnings, and explain why many workers ignore safety signs.

12.1 Symbols

Symbols are often used in industry to identify controls, machine functions and states of processes. Symbols are also widely used as traffic signs and for public information at airports and train stations. The idea is that a picture can convey many words, so a symbol can be more succinct than a label that contains many words. The other (assumed) advantage is that symbols do not have to be translated, i.e. they can be understood by individuals throughout the world. In fact, as exemplified below, many international machine manufacturers prefer to use symbols, since labels would have to be translated to the local language. But many symbols are difficult to understand, particularly those that refer to abstract machine functions that are hard to visualize or think of. It is then often better to use a label (Collins and Lerner, 1983).

12.1.1 Example: Standardization of Symbols

In 1981 the author participated in a meeting organized by the Society of Automotive Engineers. The purpose was to standardize symbols for 'off-the-road vehicles' such as construction vehicles and agricultural machines. At the very first meeting, a proposal was made to standardize no less than 134 different symbols, all of which had different meanings. As the only ergonomist in the group, I asked if there was any information on whether the symbols would be easy to understand by the users. This was obviously the wrong question. The sole purpose of this group was to standardize symbols so that labels did not have to be translated. The futility of my further collaboration was obvious, and I quickly resigned. However, I brought with me several proposed symbols, some of which are depicted in Figure 12.1. These symbols were intended for use in construction vehicles, and in order to evaluate them we asked 40 US construction workers to translate the symbols into words and actions (Helander and Schurick, 1982).

A couple of the symbols were easy to understand. The two arrows representing the up and down motion of a controlled element was an

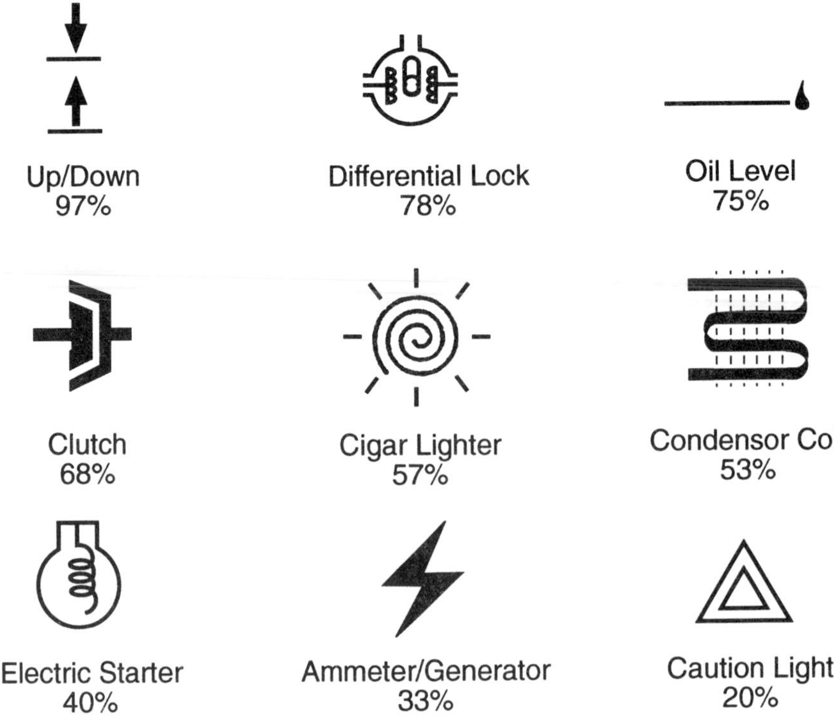

Figure 12.1 Symbols and the percentage of construction workers who understood their meaning

obvious symbol. The oil level, which was understood by 75%, is an example of composite symbols: a level and a drop of oil. The electric starter was understood by only 40%. But the worst was a caution warning which was understood only by 20%. Since this symbol is used as a traffic sign in most European countries we would assume that European construction workers would not have had much difficulty with this particular symbol.

Symbols may be more difficult to understand for individuals in industrially developing countries, due to lack of education or previous exposure to symbols. The International Standards Organization has therefore suggested that international symbols must be tested in a minimum of five different countries and they must be understood by an average of 66% of users (Zwaga and Easterby, 1982).

A large number of research studies have been performed (Lehto, 1992; Lehto and Miller, 1986). Relatively few have examined industrial symbols, but there are many studies of traffic signs, public symbols, military symbols and computer icons. The principles are fairly general and there is carryover to industrial applications.

One early study of symbolic traffic signs in Sweden investigated what percentage of drivers could recall a road sign 1 minute after passing it. Some of the results are displayed in Figure 12.2 (Johansson and Backlund, 1970). The general danger sign could be recalled only by 26%, probably because it is general and can refer to a variety of situations. The sign for frost damage is very common on Swedish rural roads. Most drivers have the experience that the damage is usually slight and will not affect driving. On the other hand, traffic signs with specific information or important information were much easier for the drivers to recall; and it is obvious that they were read. This study shows

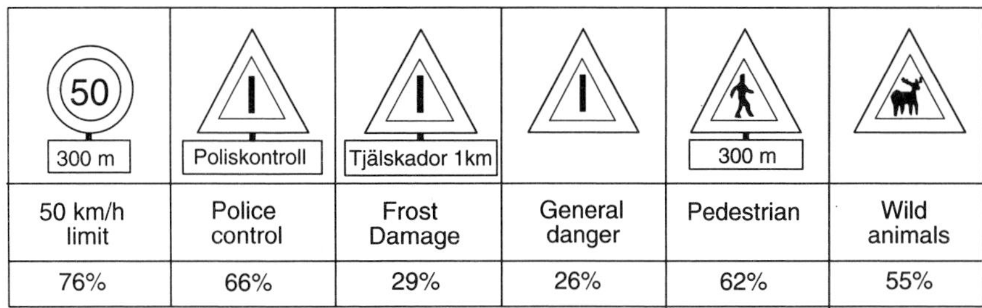

300 m	Poliskontroll	Tjälskador 1km		300 m	
50 km/h limit	Police control	Frost Damage	General danger	Pedestrian	Wild animals
76%	66%	29%	26%	62%	55%

Figure 12.2 Road signs used in a field study on a rural highway in Sweden, and the percentage of drivers who recalled having seen each sign

that individuals will remember a sign that is relevant to the situation at hand.

The results carry over to industrial situations. Generalized warning signs such as 'danger' or 'be safe' are not specific enough, as they do not instruct individuals what to do.

12.2 Labels and Written Signs

The main constraint in designing labels is that the message must be short, otherwise it will not be read (Vora *et al.*, 1994; Loewenthal and Riley, 1980). The difficulty is then to find a short message that expresses the situation and carries good semantics.

Across the university campus of SUNY Buffalo, there is a sign next to light switches carrying the following message:

Please turn out lights when not required

Most people probably read only the first line, because they understand the message anyway. But many individuals will ignore the label because it is too long. The other problem is that the message is confusing. Why would anybody turn off a light if it is not required. Much better would be the following:

Please turn off lights

Broadbent (1977) made the point that many statements can be expressed in an affirmative, a passive, or a negative manner as illustrated below.

Affirmative: 'The large lever controls the depth of the cut.'
Passive: 'The depth of the cut is controlled by the large lever.'
Negative: 'The small lever does not control the depth of the cut.'

Broadbent pointed out that an affirmative, active statement is easier to understand than a passive statement. Negative statements require a double reaction time, since the user must first understand 'what not to do' and then, only by inference, is the appropriate action clear. The same observation is also valid for traffic signs. Positive signs stating the preferred action are easier and quicker to understand than are prohibitive signs.

12.3 Warning Signs

There are several distinct stages in the information processing of warning signs (Figure 12.3). An individual is first exposed to a warning

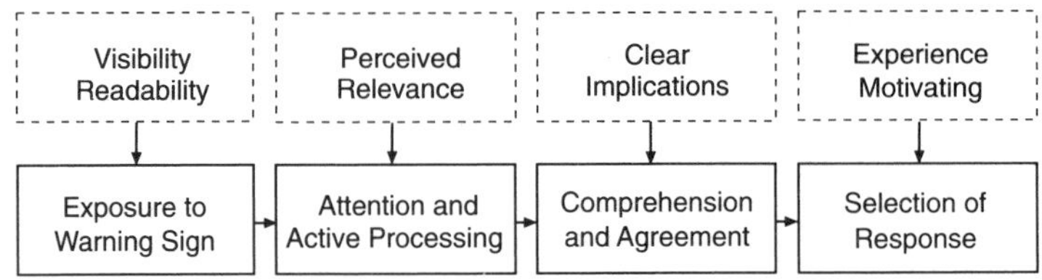

Figure 12.3 The four stages involved in the information processing of warning signs

sign. As a result there is an image of the sign on his or her retina, but the person has not yet paid attention. There are several factors that make people look at a sign (Hale and Glendon, 1987). The size of the sign is clearly important: the bigger the better. Location is also important. One should position a sign where people tend to look. Drivers, for example, usually try to look as far down the road as possible. This means that in a curve to the right they will look along the right-hand side of the road (rather than to the left-hand side) and this is also the preferred location for a traffic sign. By walking through an industrial plant one can similarly appreciate where employees look and what would be eye-catching locations for warning signs. As mentioned above, the amount of attention paid to a sign depends largely on its relevance. A sign that is not relevant to the work situation will be ignored.

Given that the worker has actively processed the sign, we will hope that the semantics of this sign will make it possible for him or her to draw a clear conclusion. The individual must then agree with the conclusion. If there is no agreement, there will be no action taken. Following an agreement, the individual must select and execute one of many alternative responses. One possible response is to do nothing, since the cost of action may be too great. This selection of response depends largely on experience, as an experienced individual has a greater response repertoire (Hale and Glendon, 1987). We discuss some of these issues in more detail below.

12.3.1 Information Processing of Warning Signs

Based on the model in Figure 12.3, we can now introduce several additional factors that are important in assessing the effectiveness of signs (Lehto, 1992).

12.3.1.1 Information Overload

Information overload is particularly obvious during driving. It is a common experience when driving in an unfamiliar city that there are too many traffic signs competing for one's attention (Lehto and Miller, 1986). Many drivers make mistakes since they do not have the time to attend to and read every sign. Often there are too many words to read on a sign, and chances are that drivers will only get halfway through the text and then try to act on incomplete information.

12.3.1.2 Attention and Active Processing

Individuals will attend to a sign if they perceive that the sign is relevant. Unfortunately, many individuals regard warning signs as irrelevant. Because the hazard is not perceived, the warning sign is not read. This is also one of the basic problems in motivating workers to work safely: the hazard is simply not perceived, so there is no need to work differently. Zimolong (1985) provided some interesting data for construction workers. Painters, who stand on ladders for much of their working time, do not perceive ladders as being unsafe. However, according to accident statistics, they are, and painters have more injuries than any other occupational group. Similarly, scaffold assemblers do not think of scaffolding as being unsafe, although 50% of all fatalities in construction work involve scaffolding.

It is possible that workers develop strong coping mechanisms that make them underestimate hazards, particularly if these hazards are inherent in their work. This being the case, it will be difficult to motivate, say, painters to consider a warning sign for ladders as being relevant. The warning contradicts the worker's mental model of what is safe and unsafe. The same basic problem prevails in safety training, which participants may think of as directed at others but not at themselves.

Another basic problem is that accidents happen fairly infrequently and, therefore, there are not enough warning examples. This is more of a problem for young workers who may never have seen an accident. Older workers are more perceptive and motivated (Dedobbeleer and Beland, 1989). Perhaps there are ways of enhancing the perceived relevance by exposing young workers to a 'benign experience' (Purswell *et al.*, 1987). Perhaps one could make something happen that could reinforce the sense of hazard. Perhaps citations written by a company safety officer would do the trick - or perhaps not. Many of these issues are not well understood, and additional research is required.

12.3.1.3 Comprehension and Agreement

In reading comprehension there is a trade-off between detailed description and the use of simple words. Simple words may not be illustrative enough, whereas detailed descriptions are not read. Wogalter *et al.* (1985) suggested that there should be four fundamental elements in a warning sign:

- Signal word: to convey the gravity of the risk, for example 'danger', 'warning' or 'caution'.
- Hazard: the nature of the hazard.
- Consequences: what is likely to happen if the warning is not followed.
- Instructions: the appropriate behaviour required to reduce the hazard.

An example of an effective minimalist warning would be:

Danger

High Voltage Wires

Can Kill

Stay Away

The main reason for a short warning is the limited capacity of the human short-term memory. Typically short-term memory can store about seven 'chunks' or concepts (Miller, 1956; Simon, 1974). But short-term memory is constantly upgraded to include current items, and half of the memory is therefore replaced in 3–4 s. Warning signs, as they are read by a driver in a car or by workers in industry, are considered only for a very short period of time. There is a quick decision to do or not to do something. The scenario is processed quickly in short-term memory, after which the situation is forgotten. Humans take short-cuts in information processing. These are necessary to sort out the important issues expediently and deal with scenarios at hand on a continuing basis. There is really no need to store such information permanently in long-term memory.

12.3.1.4 Selecting and Performing a Response

An individual may have fully comprehended a warning sign and may also be in full agreement, but may select to do something different because there is a 'cost of compliance'. For example, the decision to press the stop button on an industrial robot is countered by the realization that it may take an hour to start up the process again (Helander, 1990). Employees often select not to be safe. Safety glasses, steel-toed boots, respirators and other personal protective equipment are perceived of as being inconvenient and uncomfortable (Krohn *et al.*, 1984; Hickling, 1985). The cost of compliance is too high, unless of course the company decides to change the cost equation by enforcing safety rules. Some manufacturing companies have acquired a reputation for doing this very efficiently.

Another issue is whether the action implied by a warning sign can be incorporated in the regular work task. For most drivers, safety behaviour has become an integral part of driving. For example, drivers stop at a junction with stop signs without giving it too much thought. 'Stopping' has become an act that is totally integrated into the driving task. Could it be that safe behaviour can be incorporated equally well in a regular work task? Probably not - unless the novice worker is coached extensively by a qualified trainer (supervisor or peer), the corresponding situation to driver training on the shop floor.

Chapter 13

Development of Training Programmes and Skill Development

Due to the introduction of computers and automated tools, manufacturing and production systems have become increasingly complex (Drury, 1991). Although the automation of manufacturing systems may have the effect of removing some employees from the shop floor, those employees who remain have greater responsibilities. They supervise a production system, they participate in the planning and scheduling of production, they exercise quality control, and sometimes they take responsibility for ordering and deliveries. In this complex environment, human errors can have important consequences.

In referring to 'human errors' we do not mean to imply that an operator was truly at fault and should be blamed. Rather, the reason for error is often found in the design of the production environment, or it may be due to lack of training. In an unexpected crisis situation, there will not be much time to act, and the human operator will use his or her intuition to make decisions. The decision may be entirely logical and rational, except that sometimes the production system is not designed for logical or rational input, so a 'human' error is committed. The issue of *transfer of training* is often brought up in this context. In situations of emergency individuals act according to previously learned stereotypes – except that the old stereotypes may not fit a new environment. One example is a pilot who can fly several aeroplanes. In an emergency situation, the pilot may decide on instinct. The response was appropriate for the first aircraft the pilot learned to fly, but not the present aircraft. In previous chapters we have mostly addressed issues of design, here we focus on the training aspects of the problem.

13.1 Establishing the Need for Training

To establish the need for training one must consider both the type of task and the characteristics of the trainees (Goldstein, 1980; Patrick, 1992). This is traditionally done in a systems approach to defining training needs (Figure 13.1).

First the training objectives are defined. These explain why the training is necessary. Obviously the training should be based on perceived needs for training, and these needs must be clarified and formalized. From the training objectives 'criterion measures' are developed, which express how well the students should master the course afterwards. The training programme must be effective in improving the trainees. But there is no way of knowing whether this is the case unless the effectiveness of the training programme is evaluated after the training has taken place. The criterion measures are used for such evaluation, and they should be stated in terms of

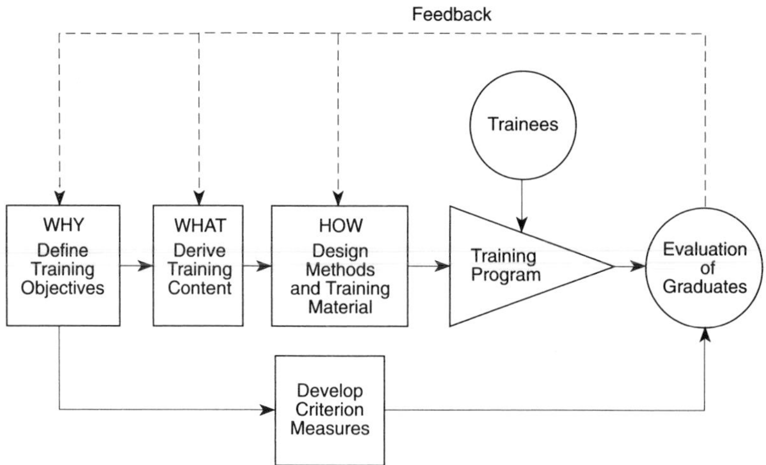

Figure 13.1 The development and evaluation of a training programme (adapted from Eckstrand, 1964)

concrete goals that can be measured after the training course. Examples of such criteria are: for manufacturing assembly - being able to assemble a widget in 5 min (skill based); for safety - being able to specify the safety procedures for several scenarios (rule based); and for manufacturing production - being able to produce a manufacturing production schedule for a mix of orders (knowledge based) (Rasmussen, 1986).

After the training objectives and the criterion measures have been clarified, the content of the training must be defined. For the skill-based and rule-based scenario a task analysis is often used. For the knowledge-based situation, where individuals have to think deeply about a problem, a traditional task analysis may not be so useful because there are usually several alternative ways for an individual to solve a problem. Task analysis is best used to portray a procedural or hierarchical task.

13.2 Determining Training Content and Training Methods

The type of training depends on the type of task. In Table 13.1 we use Rasmussen's (1986) distinction between tasks that are skill-based, rule-based, or knowledge-based. The skill-based scenario refers to manual or perceptual motor skills (e.g. driving). Over time these skills become so well learned that they become automatic. Rule-based skills are used for procedural tasks. Here the worker must first recognize

Table 13.1 Skills and options for training

Type of skill	Example of tasks	Training options
Skill based	Manual skills: assembly	On the job, coaching
Rule based	Procedures: workplace organization, housekeeping, safety procedures	On the job, coaching, classroom
Knowledge based	Problem solving: production scheduling	Classroom, problem solving at work

a scenario and then decide what to do. For example: 'If Condition A, then Action 1'; 'If Condition B, then Action 2'; etc. The knowledge-based tasks are those that require deep thinking about and pondering on alternatives such as in problem solving.

As we suggest in Table 13.1 these skills should be trained differently. On-the-job training is clearly appropriate for skill-based jobs (Holding, 1986). The job site has better ecological validity than the classroom. At the job one finds the real scenarios, which remain difficult to simulate in a classroom. Classroom training is appropriate for rule-based and, particularly, knowledge-based jobs. The knowledge-based scenario has many theoretical elements (knowledge of mathematics and physics) which may be useful in problem solving.

The options for training listed in Table 13.1 are quite traditional. Many researchers have, in fact, expressed reservations against the use of fashionable training tools such as computer-aided instruction and interactive TV systems. There is nothing wrong in using such tools, but the availability of tools should not dictate the strategy for training, and this has often been the case. As Gilbert (1974) expressed:

> If you don't have a gadget called a teaching machine don't get one. Don't buy one; don't borrow one; don't steal one. If you have such a gadget get rid of it. Don't give it away, for someone else might use it. This is a most practical rule based on empirical facts from considerable observation. If you begin with a device of any kind you will try to develop the teaching program to fit that device.

13.3 The 'Why?', 'What?' and 'How?' of Training Development

As illustrated in Table 13.2, there are two basic scenarios: training to learn a new job, and training to improve job performance. In a manufacturing plant, workers, supervisors and management may assess the need for training by analysing task requirements. For existing production systems one can analyse production reports and quality reports, since these may give important hints of what to train. There are three stages in training development: objectives, content

Table 13.2 The WHY? WHAT? and HOW? of training development

Development of:	Training	
Objectives Content Materials	To learn a new job	To improve job performance in an existing job
WHY? Define training objectives		
• Production reports		x
• Quality reports		x
• Customer feedback		x
• Employee feedback		x
WHAT? Derive training content		
• Employee information	x	x
• Experience with similar case	x	x
• Expert option	x	x
HOW? Design methods and training materials		
• Task analysis	x	x
• Discussion with employees	x	x
• Experience with similar case	x	x

and design corresponding to the three questions 'Why?', 'What?' and 'How?' in Table 13.2.

In the first stage ('Why?'), training needs are diagnosed by using feedback from customers and employees and reports of production, quality and yield. Note here that feedback requires an existing scenario. For new jobs there is no direct feedback, since there are no production reports or customer complaints.

The analysis of new jobs starts at the second stage ('What?'). For a new scenario the analysis must be based on past experience (as this may apply), conventions, and trial and error. Feedback is missing unless one can draw from experience of similar cases within the company, or from the outside expert opinion.

In the third stage ('How?'), the training programme is designed in detail. These details are based on familiarity with the work scenario, and are derived from task analyses and discussions with employees.

13.4 Use of Task Analysis

Task analysis has been used extensively as a tool to develop training programmes (Drury, 1983). The main purpose of task analysis is to obtain a thorough understanding of the task, and thereby capture what is important training. Task analysis has several additional purposes, such as the design of safety systems and the design of workplaces. We do not give examples of such analyses here, but the procedures involved are quite similar to those illustrated below.

In task analysis a job is first broken down into its various components. The US Air Force has suggested a hierarchical breakdown:

- Job
- Duty
- Task
- Sub-task
- Activity

The purpose of the hierarchical breakdown is to provide a logical description of the various activities that constitute the job; an example is given in Figure 13.2. In this example, the job of a car mechanic can be regarded as being composed of three major duties: relining the brakes, tuning the engine, and servicing the cooling system. These duties are broken down into tasks. In tuning an engine there are three major tasks: repair the distributor, replace the plugs, and repair the carburettor. For some of these tasks a further breakdown of activities may be necessary, e.g. remove the distributor cap and replace the points.

The extent to which a task is broken down depends on the purpose of the task analysis; one could stop at the activity level, or one could go further and study very minute details, such as eye movement and finger movements. This is not warranted for training development. But in designing a cockpit it is useful to analyse eye movements because such information can suggest where displays should be located.

Depending upon the purpose of the analysis, the analyst will have to define what information needs to be collected and how best to present the information so that it is useful. There are some common tools used to present task-analysis information. The most common is a table composed of columns (Table 13.3). These particular tables were developed to describe the task of a process control operator

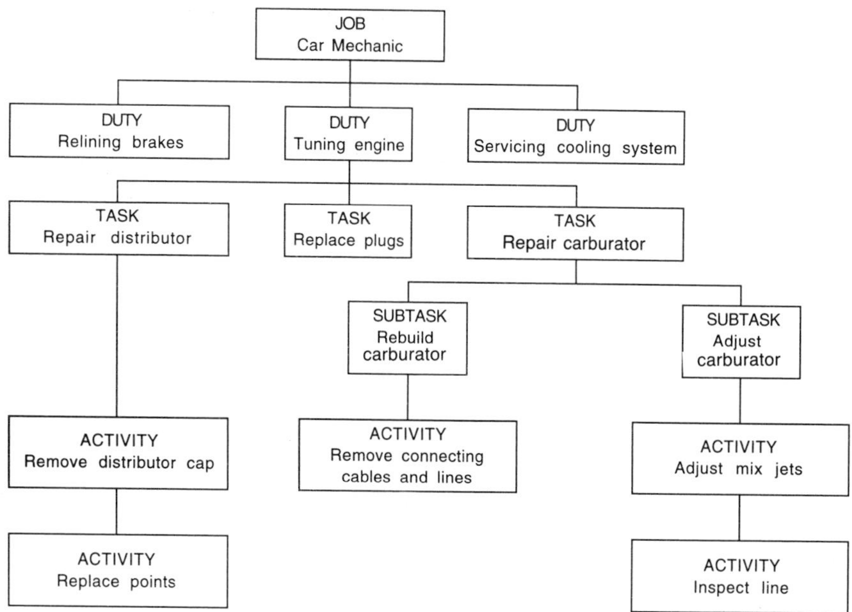

Figure 13.2 An example illustrating the hierarchical breakdown of a task (US Air Force)

Table 13.3 Two different column formats used to describe the task of a process control operator

Task	1. Hardware interface, software interface	2. Visual and manual cues	3. Required response	4. Operator feedback	5. Performance criteria, e.g. time and accuracy	6. Criticality of action
1						
1.1						
1.2						
2						
2.1						

TASK DESCRIPTION					TASK ANALYSIS					
Task or step number	Instrument or control	Activity	Cue for initiation	Remarks	Scanning perceptual anticipatory requirements	Recall requirements	Interpreting requirements	Manipulative requirements	Likely human error
1									
1.1									
1.2									
2									
2.1									

(Drury, 1983; Swain and Guttman, 1980). The definition of the column headings and the description of the task is up to the analyst. For every analysis one must define what information is necessary and how to present it, so that it is useful for design. There are no set formats or general procedures for a task analysis – it is an art, not a science (Montemerlo and Eddower, 1978).

The training programme is defined based on the results of the task analysis. Some aspects of a task will be more important to train than others, and in Table 13.3 the criticality of operators' actions is analysed. Although it is important to train the entire task, actions that are highly critical should be emphasized.

After the completion of training it is important to evaluate the effectiveness of the training, using the criterion measures developed previously. The results of these evaluations can then be fed back to refine the training objectives, training content and modify the training methods (Figure 13.1).

13.5 Training in Manufacturing Skills

Manual assembly is highly procedural and is easier to describe by using task analysis than are many other jobs. Training of manual skills is best performed 'on the job' at the workstation. This does not imply that training should be haphazard. On-the-job training can be highly structured. There are some important guidelines for training in manual skills. In particular, *providing feedback on the quality of work*. Feedback, also referred to as 'knowledge of results', is particularly important during the first few days on a new task (Salmoni *et al.*, 1984). An effective coach will point out good and bad aspects of task performance. With feedback an individual will reach the maximum performance level much faster (Figure 13.3). These considerations are particularly useful in small-batch manufacturing, since the production of each batch is limited, and it is important to reach a high level of proficiency quickly.

13.6 Part-task versus Whole-task Training

The main advantage in training subtasks rather than the whole task is that subtasks are easier to learn. After one subtask has been trained one can continue training similar tasks, thereby using the transfer-of-training effects. The disadvantage is that one may lose dynamic or contextual cues that are available only if the entire job is trained at the same time (Fitts and Posner, 1973; Holding, 1986). The transition between subtasks becomes difficult. Think of a pianist learning 10 bars of music at a time – the interpretation and dynamics of the entire piece of music would suffer.

When should one use part-task and when should one use whole-task training? Researchers have not yet come to a conclusion. As often is the case in ergonomics, we cannot give a general answer other than: 'It depends on the task'.

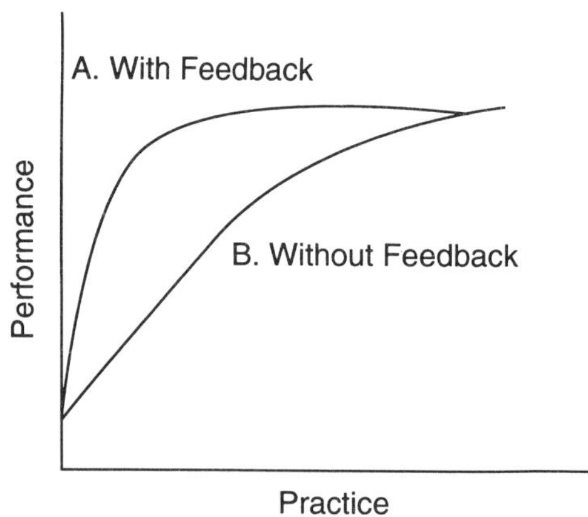

Figure 13.3 Training with feedback improves job performance significantly

13.7 Use of Job Aids

Job aids are used in training, as well as later when the operator has learnt the task. Sometimes there are things to memorize, which are difficult to train, but still necessary for the task. Most of us have used 'cheat-sheets' - small pieces of paper which are pasted to a VDT to help us remember command names. Such job aids can be critical, as the following example illustrates.

13.7.1 Example: Remembering Error Codes

The author was once called to analyse the ergonomic design of several microscope workstations. The microscopes were used for quality control in the manufacturing of electronic boards. Management had anticipated recommendations to improve the seating posture, the illumination of the workstation and the microscope itself. However, there was a much more important problem. The operators' primary duty was to report quality defects of components, and there were 24 different kinds of defect. For each defect that was found the operators had to write an eight-number error code on a sheet of paper which accompanied the board. In case the error code had been forgotten, operators could find it in a 250-page manual. This was an awkward procedure, and operators frequently took a chance, hoping to report the correct error code.

To help operators remember the error codes we designed a job aid. We 'condensed' the manual to a plastic cube with 8 cm sides. On each of the six sides, four quality defects were illustrated with a figure and the corresponding error code. This job aid simplified the task significantly.

There are many types of job aid available. In discussing the use of computer manuals, Carroll (1993) suggested that manuals are rarely read. He advocated the use of a *minimal manual*, which consists of only a few pages. The point is that job aids must be easy to use. There are many possible formats that can be used for job aids (Kinkade and Wheaton, 1972).

- *Procedural instructions, flowcharts, tables, and codebooks*. The main concern here is to condense the information into a format that takes the minimum space, and yet is instructive.
- *Colour coding* is sometimes used in a workstation to connect parts or procedures that belong together by using the same colour.
- *Schematic diagrams* and *graphics* are often used in process control tasks. Sometimes they are painted on the control panel to suggest causal relationships.
- *Checklists* are used by an aeroplane pilot to follow the necessary procedure in taking off and landing an aircraft. Checklists are likewise used in industry to help people remember long procedural tasks.
- *Computer help systems* can be made available for any computerized task. Help functions must be easy to access and easy to understand. Most of us have bitter experience of computer help systems that are impossible to use. Ergonomics expertise is necessary in designing usable help systems.

13.7.2 Example: Study of Job Aids

The US Navy has taken much interest in the design of job aids for maintenance. As rumours go, at any time about 40% of the systems on a naval ship are in need of maintenance. The systems are very complex and require great skills to maintain. A study by Elliott and Joyce (1971) demonstrated the importance of a procedural job aid which was used in locating faults in Doppler radar systems. In addition to the procedural job aids there were a dozen repair manuals which had to be used to accomplish the task. The main function of the procedural aid was to find one's way in the manuals. An experimental task was performed using two groups of subjects:

(A) High-school students with procedural aids but no previous experience of the task.
(B) Maintenance technicians with 3–6 years experience of the task but no procedural aid.

It turned out that many of the high-school students performed better than the technicians. As an example, one skilled technician took 12 h to find the problem in a Doppler radar system, performing 133 steps and referring to 41 sections in eight separate manuals.

One high-school student using procedural aids for the same task took 5 h performing 35 steps and referring to three sections in three manuals. From this study it is obvious that job aids can be of very great importance in improving task performance.

13.8 The Power Law of Practice

Practice makes perfect – but how much time will it take? The power law of practice can actually predict future performance times. This law is well known to industrial engineers who have knowledge of time-and-motion studies. It is well illustrated in a classic study reported by Crossman (1959). Crossman obtained performance records for a woman in Tampa, Florida, who was working in a tobacco factory rolling cigars. The improvement in performance over a time period of 7 years is illustrated in Figure 13.4.

During the first year the woman rolled approximately 1 million cigars and her performance time was getting close to the machine cycle time, but even after 7 years and 10 million cigars there were still improvements in performance. The results suggest that the performance of complex manual skills will continue to improve over time. For the woman in Tampa, the machine cycle time set a limit to how much she could improve.

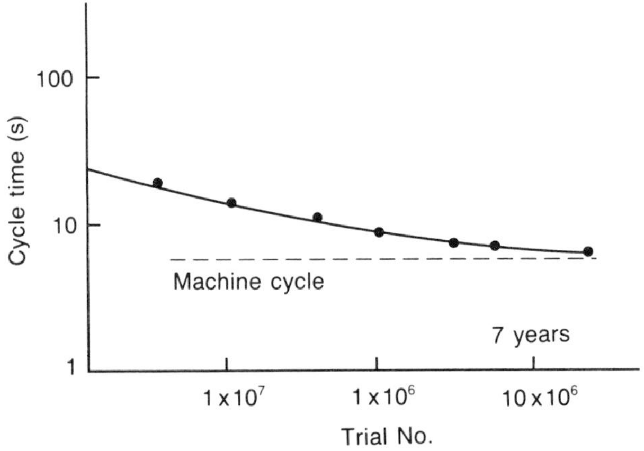

Figure 13.4 Improvement in performance time for rolling cigars

The relationship between performance time (T_N) and the number of trials follows the power law of practice (Welford, 1968; Konz, 1990):

$$T_N = T_1 N^{-a}$$

where T_1 is the performance time on the first occasion, T_N is the performance time on the Nth occasion, and a is a constant. Rewritten in logarithmic form, we obtain a linear relationship (Figure 13.5):

$$\log T_n = \log T_1 - a \log N$$

The power law of practice is used to predict the performance times for industrial tasks; an example is given below.

13.8.1 Example: Prediction of Future Assembly Time

A company is trying to estimate how the assembly time of a machine part will decrease in the future when the operator has more experience. They plan to use the predicted assembly time to estimate future manufacturing costs. The assembly time of the operator was measured. For the 1000th trial the assembly time was 121.8 s and for the 2000th trial the assembly time was 11.5 s. What is the expected assembly time for the 50 000th trial?

Solution: from the equation above two expressions of T_1 are obtained:

$$T_{1000} \times 1000^a = T_{2000} \times 2000^a$$

$$a = 0.36$$

$$T_1 = T_{1000} \times 1000^{-0.36} = 18 \text{ s}$$

$$T_{50\,000} = 18 \times 50\,000^{-0.36} = 3.7 \text{ s}$$

The improvements in assembly time are substantial and the manufacturing costs can be predicted to decrease accordingly.

The slope of the learning curve (as in Figure 13.5) is used by industrial engineers to express the rate of improvement, or the learning rate. The learning rate is expressed as a percentage and is obtained by dividing the performance times for the 2Nth trial by the Nth trial. In our example the learning rate can be calculated by dividing the

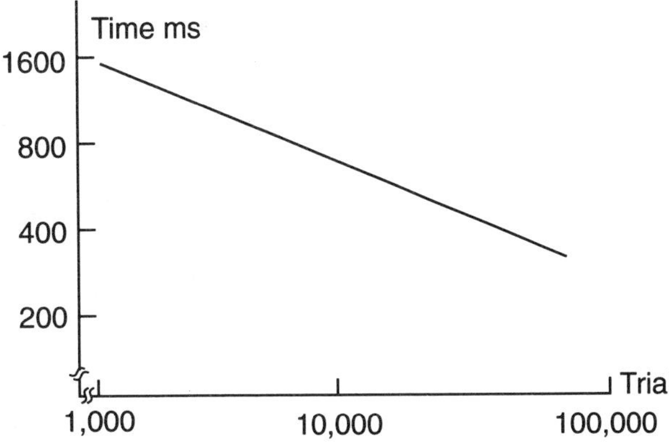

Figure 13.5 Performance time decreases with the number of trials - when plotted in a log–log diagram, the relationship is linear

assembly time of the 2000th trial by the assembly time for the 1000th trial, and we obtain:

$$\text{Learning rate} = \frac{11.5}{14.5} = 78\%$$

Konz (1990) reported learning rates for several different types of task, and part of this information is reproduced in Table 13.4. Two important issues are illustrated in this table.

1. *Complex tasks improve more than simple tasks*. For example, from the table we note that truck-body assembly improves much more (68%) than a simpler task such as attending a punch press (95%). This principle is also illustrated in predetermined time systems that are used to predict assembly time. For example, in MTM-1 motions are broken down into 10 categories: reach, move, turn, apply pressure, grasp, position, release, disengage, body motions and eye motions. For the very simple types of motion such as 'reach' and 'move' there is very little improvement over time. These are primitive movements of the hand back and forth, and there is not much that can be improved. For more complex movements that require manual skills, such as 'positioning', there is more opportunity for improvement. 'Positioning' typically involves intricate assembly movements for alignment and orientation of parts.
2. *There is not one 'correct' learning rate*. For a task such as grinding, Table 13.4 gives two different learning rates (82% and 98.5%). This illustrates that grinding tasks can be very different; some of them involve complex movements for which there is a greater potential for improvement.

13.9 Recommended Reading

For more information about the use of predetermined time systems, we refer the reader to two texts. Konz (1990) gives a good account of MTM, and Zandin (1990) presents MOST, which is a more recent system.

Table 13.4 Learning rates for several types of task (adapted from Konz, 1990): the lower the learning rate, the greater the improvement in performance time

Learning rate (%)	Type of task
68	Truck body assembly
74	Machining and fitting small castings
80	Precision bench assembly
82	Grinding
83	Servicing automatic transfer machines
84	Cigar making
88	Welding (manual)
89	Punch press, milling
90	Punch press
92	Assembly with jig, welding
95	Punch press, screwdriver work
95	Word-class mail runner
98.5	Grinding, milling, assembly

Chapter 14

Noise

Noise is very physical and noticeable to most employees. Questionnaire investigations in industrial plants show that workers usually single out noise as the most important problem (Karlsson, 1989). This is not totally unexpected because, compared with many other ergonomic problems, noise is very obvious and concrete.

In this chapter we first discuss several different methods for assessing the effects of exposure to noise. We then examine some performance effects of noise that are likely to affect an industrial worker, and we will discuss engineering methods for reducing noise in the workplace. Table 14.1 gives some examples of typical noise levels.

14.1 Measurement of Sound

Sound-level meters are used to measure sound. They consist of a microphone, an amplifier, and a meter that gives a visible reading in decibels (dB) on a scale. Most meters incorporate three different types of weighting of the sound. These are known as the A, B and C scales (Figure 14.1). In particular, the dBA scale has achieved widespread use in most industries. This scale (or weighting function) approximates the sensitivity of the human ear. The dBA scale is referenced to a sound pressure level of 0.00002 N m^{-2}, which corresponds to the threshold of hearing.

To calculate the sound pressure level (L_p, in decibels) the following formula can be used:

$$L_p = 20 \log \frac{P}{P_0}$$

where P is the root mean square (r.m.s.) sound pressure, and P_0 is the reference sound pressure (0.00002 N m^{-2}. From the formula it

Table 14.1 Examples of activities and corresponding noise levels

Activity	Typical noise level (dB)
Near jet aircraft at take-off	125
Punch press at 1 m	105
Lathe	90
Quiet manufacturing (e.g. electronics)	75
Automobile at 20 m	65
Conversion at 1 m	50
Inside quiet home	42
Public library	20
Recording studio (threshold of hearing)	0

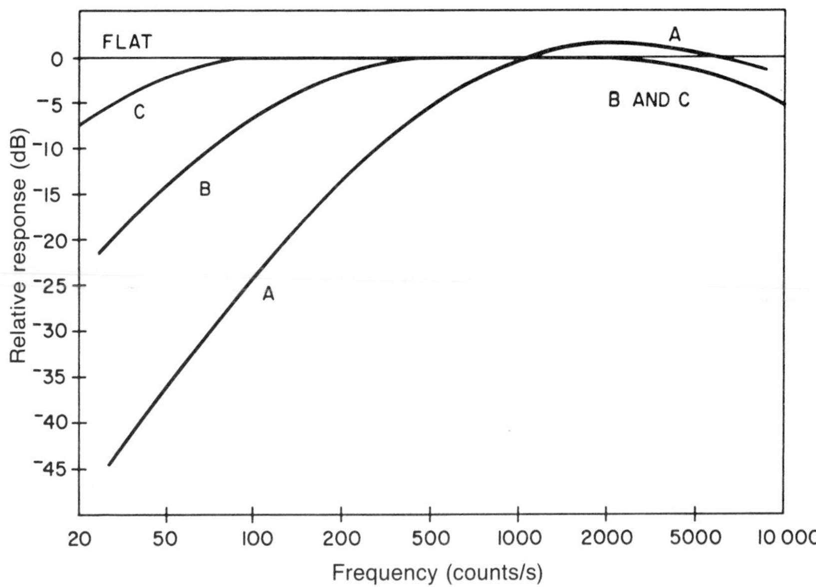

Figure 14.1 Weighting curves A, B and C for sound-level meters. A is less responsive to lower frequencies and gives the best approximation of the sensitivity of the human ear

can be derived that doubling the sound pressure would lead to an increase of 6 dB.

In most countries there are laws that regulate how much noise employees can be exposed to. In the USA the maximum noise exposure throughout a working day of 8 hours is 85 dBA. For every 5 dBA increase beyond 85 dBA the exposure time is reduced by half, e.g. if the noise is 90 dBA then the maximum exposure time is 4 hours, and for 95 dBA it is 2 hours. According to the Occupational Safety and Health Administration (OSHA) regulations in USA, noise exposure of different intensity can be added according to the formula:

$$D = \sum_i \frac{C_i}{T_i}$$

where D is the allowable noise dose (should be ≤ 1), C_i is the number of hours of exposure to a noise level i, and T_i is the permissible number of hours of exposure to noise level i.

14.1.1 Example: Calculation of Noise Dose

A machine subjects its operator to 85 dBA when it is idle and to 90 dBA when it is used at full power. Assume 7 hours of use per day, with 2.1 hours at 85 dBA and 4.9 hours at 90 dBA. The total noise dose is calculated accordingly:

$$D = \frac{2.1}{8} + \frac{4.9}{4} = 1.487$$

Since the noise dose is greater than 1.0, this work exposes its operator to excessive noise that is not permissible.

There are four different aspects that can make noise unacceptable in the working environment.

1. Noise can cause hearing loss.
2. Noise can affect performance and productivity.
3. Noise can be annoying.
4. Noise can interfere with spoken communication.

14.2 Noise Exposure and Hearing Loss

The major concern about noise exposure is loss of hearing. There are two major types of hearing loss: conductive hearing loss and neural hearing loss. *Conductive hearing loss* can be caused by mechanical rupture and/or dislocation of the eardrum and the bones in the middle ear (Figure 14.2). This may be due to a sudden intense pressure wave, such as produced by an explosion or a blow to the external ear. As a result there may be physical damage to the middle ear, for example by dislocation of the stirrup. The hearing loss may be partial or total, temporary or permanent (Kryter, 1985).

Prolonged noise exposure can cause hearing loss due to auditory nerve damage, also called *neural hearing loss*. In this case the intensity, frequency, and duration of exposure must be considered. For example, noise levels at about 130 dB may cause swelling of the hair cells of the organ of Corti on short exposure and destruction of these cells on longer exposure. These changes are usually localized and involve only part of the organ of Corti, corresponding to certain (high) frequencies. The destruction of the hair cells in the organ of Corti is an irreversible process and the resultant hearing loss is permanent. However, if the noise exposure time is short there may be only temporary swelling of the organ, which is reversible and causes only temporary hearing impairment, called temporary threshold shift (TTS) (Ward, 1976).

A person with auditory nerve damage first loses hearing of the higher frequencies at around 4000 Hz (Loeb, 1986). It is then difficult to hear a woman's voice but it is easier to hear a man's lower pitch. The person affected soon begins to speak louder and in a monotone voice, since the modulating effect of hearing is impaired. Because low tones are

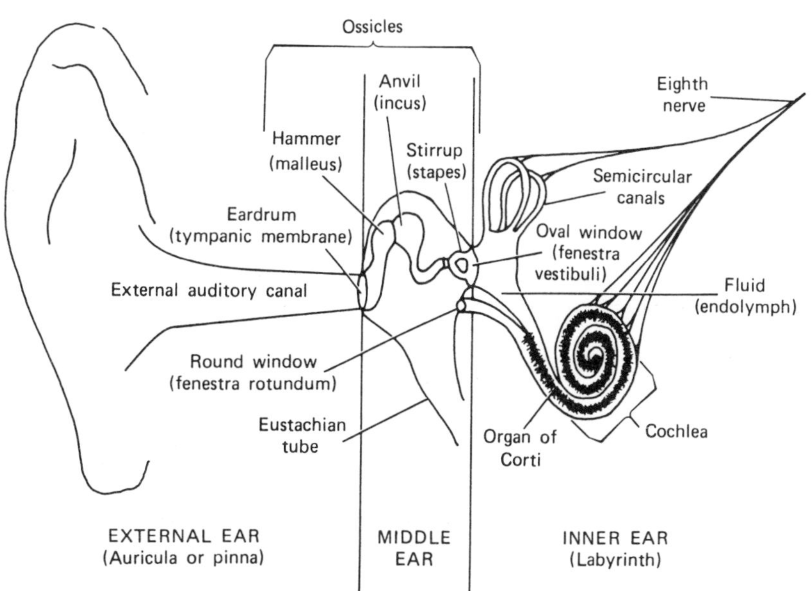

Figure 14.2 The structure of the ear. Some of the mechanisms are exaggerated in size to illustrate their functionality

heard better than higher ones, it becomes difficult to understand words and sentences. Low pitch noise seems unduly loud and conversation becomes difficult in a noisy environment. Amplification of the sound through a hearing aid may not solve any problems, since high frequencies will still not be heard.

In contrast, a person with conductive deafness will complain that others in conversations do not speak loud enough. Understanding is not impaired if the sound level is sufficiently high, and this person can hence profit from the use of a hearing aid.

As mentioned, the first and most notable damage caused by excessive noise is to hearing frequencies at about 4000 Hz. However, there is extreme variability in the individual reactions to noise. Similar loss of hearing may also occur because of ageing (presbycusis). To complicate things further, loss of hearing is also caused by ear infections, several diseases (mumps, measles, scarlet fever), and by common colds. Helander (1992) suggested that presbycusis may actually be caused by the cumulative effect of common colds over a lifetime. These viral infections can destroy auditory nerve cells.

14.3 Hearing Protectors

There are two types of hearing protector that are commonly used in industry: ear plugs and ear muffs. The plugs are designed to occlude the ear canal and are available in many types of material. Cotton has traditionally been used but, unfortunately, and contrary to popular belief, it affords no protection (Zenz, 1981). Ear plugs made out of rubber, neoprene, glass down and plastics offer good protection. Custom moulded ear plugs are also available (Casali and Park, 1990), which are individually made to fit the ear canal, and can offer excellent protection and last for a long time. Ear muffs are designed to cover the entire external ear. They consist of ear cushions made of soft spongy material or specially filled pads to ensure a snug fit.

Ear plugs provide a sound attenuation of between 15 dB for low frequency sounds and 35 dB for higher frequencies. At frequencies above 1000 Hz muffs provide about the same protection as plugs. At frequencies below 1000 Hz certain muffs provide more protection than plugs. Ear plugs and ear muffs may be worn together in intense noise situations. This combination provides an additional attenuation of approximately 5 dB.

Workers who regularly wear ear protection report that they actually hear conversations better. Cutting down the noise level reaching the ear decreases the distortion in the ear so that speech and warning signals are actually heard more clearly. An analogy can be drawn with wearing sun glasses to reduce excess glare and thus improve vision.

14.4 Analysis and Reduction of Noise

There are two main methods for measuring noise: use of dosimeters and sound level meters. Workers' exposure to noise can be quantified using a noise dosimeter (ANSI, 1991). A dosimeter is attached to the worker's body, e.g. on the chest. It summarizes the noise exposure over one working day, and it is therefore possible to assess whether an individual, with his or her particular work habits, has been overexposed to noise.

The other way is to use a sound level meter to analyse the working environment and obtain readings of the noise produced by various machines. A sound level meter can be set at different frequencies and a curve is constructed (Figure 14.3).

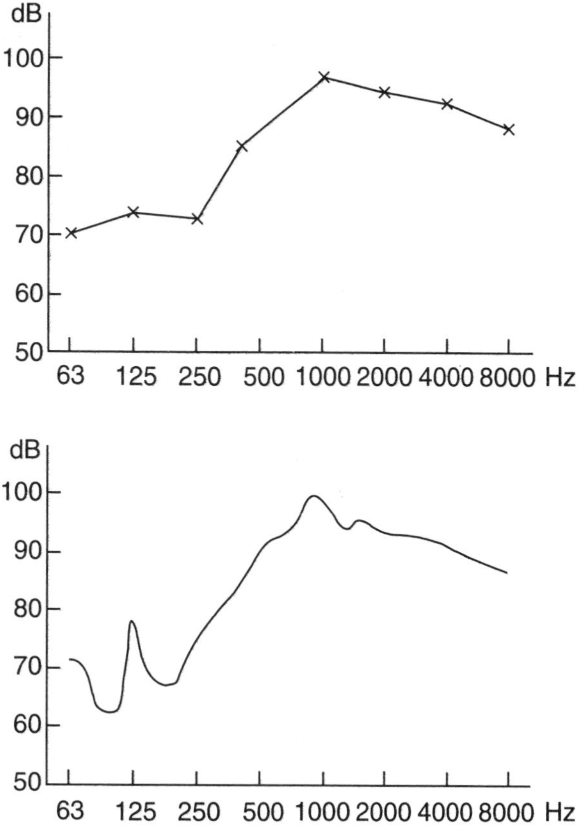

Figure 14.3 Octave-band and third-octave-band analyses for a wood planer machine. Note how the increased resolution makes it possible to identify 125 Hz and 1000 Hz as critical frequencies

There are two common types of analysis: octave-band analysis and third-octave-band analysis. In an octave-based analysis the noise is measured at each octave. The preferred practice is to divide the audible range into 10 bands having the central frequencies 31.5, 63, 125, 250, 500, 1000, 2000, 4000, 8000 and 16 000 Hz (ANSI, 1986). However, this may not give sufficient resolution for a detailed analysis of the noise. With a third-octave-band analysis there are three readings for every octave, which increases the resolution considerably. Figure 14.3 illustrates the frequency spectrum of a wood planer machine, where the noise spectrum was recorded using both an octave-band and a third-octave-band analysis. There are two peaks at about 125 Hz and about 1000 Hz. These peaks are due to the rotating elements in the wood planer machine. Since we now understand where the noise comes from, it is possible to take engineering measures to reduce these two peaks.

Through engineering change, noise energy can sometimes be moved in frequency to solve a noise problem, as has been shown by the US Department of Labor (1980). A large diesel engine in a ship was designed to operate at a 125 revolutions per minute (r.p.m.) with a direct drive connection to the ship's propeller. Noise of 125 Hz from the propeller would have been extremely disturbing to the crew. The solution was to add a differential gear between the engine and propeller in order to gear down the propeller's speed to 75 r.p.m. A larger propeller was also required. Shifting the noise to a lower mostly inaudible frequency resulted in much less disturbance.

Only a detailed analysis of the frequency spectrum of the noise source can reveal such possibilities. In the case of the wood planer it may be possible to modify the cutting speed of the machine, which could possibly reduce the noise level to a legal level of 85 dBA.

14.4.1 Reduction of Noise in Manufacturing Plant

In a manufacturing plant one can take many different measures to reduce the noise (US Department of Labor, 1980). The noise can be controlled at the noise source, by reducing the structure-borne transmissions of noise, and by reducing the air-borne transmissions of noise. A summary of some common measures is given in Table 14.2.

Many of the common noise sources in a plant (from manufacturing processes and machinery, air intake and other equipment) are illustrated in Figure 14.4. Several measures have been taken to reduce noise, including: vibration isolation mounts, placing heavy vibrating equipment on a separate rigid structure, and the use of an air intake muffler with laminar flow of air. The structure-borne transmissions have been reduced, for example by use of flexible pipe on the air intake, and sound isolating joints between the vibrating equipment and the floor. Finally the air-borne transmission has been reduced by use of sound absorbing ceilings and shields, and by enclosing noise sources in a control room and in the basement.

Most of the engineering measures listed in Table 14.2 are equally effective in reducing vibrations as well as noise. In fact, vibrations and noise are concomitant; noise and sound are in fact vibrations of the air mass introduced by compressions and rarefactions of the density of air molecules. A vibrating plate will vibrate air masses and produce noise.

14.5 Effects of Noise on Performance

There are no clear-cut effects of noise on performance. In fact, this has been a much debated topic among researchers (Broadbent, 1978; Poulton, 1978; Kryter, 1985). Gawron (1982) reviewed 58 noise

Table 14.2 Approaches to reducing noise

Control of	Measures
Noise source	Use vibration isolation mounts Fasten members to rigid structures Use mufflers on exhaust/intake Change direction of sound emission Reduce the radiating or vibrating efficiency of sound sources, e.g. by drilling holes in plates or covers
Structure-borne transmissions	Decouple source from transmitting solid Isolate using spring steel or rubber plate Use flexible couplings on shafts Use damping materials in ducts and conveyors
Air-borne transmission	Increase distance between source and worker Rotate noise source Use barriers and baffles Enclose noise source and/or workers Apply damping material Use ear protection

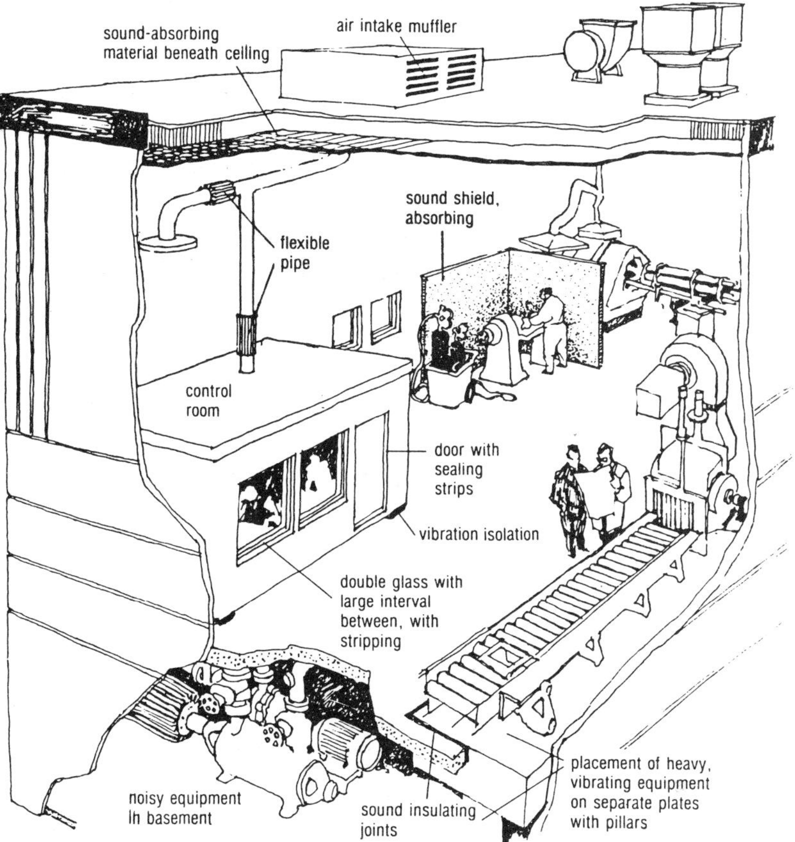

Figure 14.4 Example of noise control measures that can be implemented in an industrial building (US Department of Labor, 1980)

experiments and found that 29 showed a reduction in performance, 22 showed no effect, and 7 showed that noise improved task performance. Part of the problem in research is to provide a theory of the effects of noise on performance. If a viable theory were at hand, experiments could be undertaken and the theory tested.

A problem in formulating a theory is that there are many types of noise and many types of task. Noise can be anything from intermittent to continuous and from music to white noise. The task can be skill based (manual automatic behaviour), rule based (if scenario A, then do X; if B, then Y, etc.), or knowledge based (requires deep thinking and pondering of alternatives) (Rasmussen, 1986). The existing studies simply do not cover a sufficient range of noise and task conditions to be able to draw firm conclusions and formulate a viable theory.

In reviewing the literature one can draw a few guarded conclusions:

1. Visual functions, such as visual acuity, eye focusing, and eye movements are little, if at all, affected by noise.
2. Motor (manual) performance is rarely affected by noise.
3. For the performance of simple, skill-based, routine tasks, noise may have no effect.
4. For rule-based tasks where the individual makes quick choices between different alternatives, noise may have some effect, particularly if the noise is louder than 95 dBA.
5. The detrimental effects of noise seem to be associated primarily with knowledge-based tasks, where operators must apply their

knowledge of different scenarios, think hard and make tentative conclusions. This involves heavy use of the short-term as well as long-term memory, and the short-term memory capacity is likely to be exceeded. For example, processing of verbal, semantic information (which can be a knowledge-based task), suffers in noise well below 85 dBA. Weinstein (1977) reported that a 68–70 dBA noise level significantly impaired the detection of grammatical errors (knowledge-based) in a proof-reading task, but the same amount of noise did not appear to have any adverse effects on the ability to detect spelling errors (rule-based task).

14.5.1 Broadbent and Poulton's Theories

In the 1970s these two famous researchers engaged in a lively debate on the effects of noise (Broadbent, 1978; Poulton, 1978). They had very different theories about the effects of noise on performance (Table 14.3). Both researchers made reference to Yerkes–Dodson's law, which postulates an inverted U-curve relationship between stress and performance (Yerkes and Dodson, 1908; Figure 14.5).

Table 14.3. Poulton's and Broadbent's theories on the effects of noise on performance

Poulton

- Noise masks acoustic task-related cues and inner speech. People cannot 'hear what they think'
- Noise is distracting
- There is a beneficial increase in physiological arousal when noise is first introduced, but this beneficial increase lessens over time

Broadbent

- The detrimental effects of noise are due to 'overarousal' (Figure 14.5) and not to the masking of inner speech
- At high noise levels there is a funnelling of attention (due to overarousal). People cannot focus attention on a wide variety of information, but tend to 'lock' on the most important information. As a result errors are committed, but the operators may not be aware of these errors

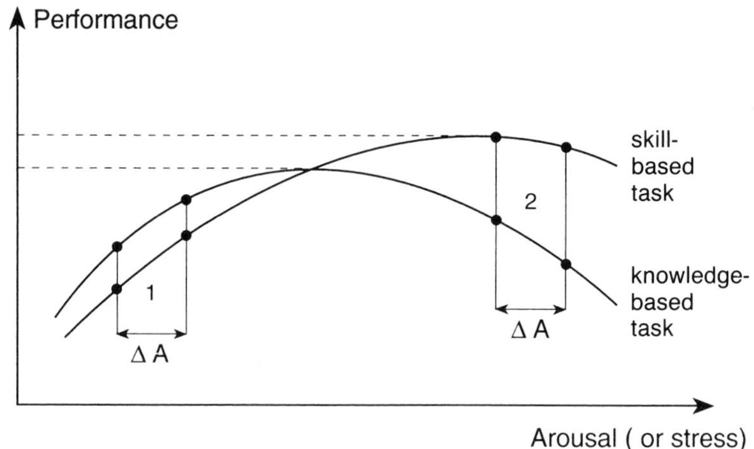

Figure 14.5 Yerkes–Dodson's law, formulated in 1908, postulates a relationship between arousal (or stress) and performance. The curves are original, but the classification of skill-based and knowledge-based tasks are those of the present author

An increase in arousal (ΔA), can have the effect of improving performance, as in situation (1) or hampering performance, as in situation (2) in Figure 14.5. There is an optimal level of arousal (or stress) at which an operator performs the best. If arousal is increased further (e.g. the task gets to be too stressful) performance will suffer. Conversely, if the arousal level is very low (a typical task with low arousal is visual inspection) then people have problems staying awake or being alert enough, and performance may then suffer due to 'underarousal'. Note that in Figure 14.5 skill-based (easy) tasks suffer less from overarousal than do knowledge-based (difficult) tasks.

So who is right – Poulton or Broadbent? As with many theories of human behaviour, the truth may have elements of both Poulton and Broadbent (Sanders and McCormick, 1993). More research will be necessary to develop the final answer to these intriguing problems.

14.5.2 Example: Discussion of Theories

Within the frameworks outlined by Poulton, Broadbent, Yerkes–Dodson and Rasmussen, discuss why:

1. Noise may facilitate certain tasks such as repetitive assembly.
2. Noise may degrade performance on tasks requiring information processing such as working on manufacturing orders, calculations of pricing, billing and shipping information.

Discuss the opposite scenario:

3. Can noise sometimes degrade repetitive assembly?
4. Can noise sometimes improve the performance in problem solving?

14.6 Annoyance of Noise and Interference with Communication

There are psychological effects of noise; reportedly people become irritated and annoyed. But the amount of irritation depends on the circumstances. Much research has gone into assessing the effect of noise (e.g. traffic noise) on communities. Sperry (1978) noted that there are many acoustic, as well as non-acoustic, factors which influence the reaction to traffic noise. Among the non-acoustic factors are: the time of day, the source of noise, and the attitude of the exposed person. The night-time tolerance level for noise is about 10 dB lower than the daytime tolerance. Noise from aircraft is perceived as more annoying than the noise from automobiles and trucks. In fact, vehicular noise needs to be about 10 dB higher than aircraft noise to be equally annoying. Finally, the attitude to noise is very important. Comparative studies have demonstrated that individuals living in Rome, Italy, tolerated a 10 dB greater noise level than did people in Stockholm, Sweden. Stiff Swedes – or *laissez faire* Italians?

Surveys in industry have shown that noise is the primary source of dissatisfaction or annoyance (Karlsson, 1989). Perhaps this is because noise is so physical and so evident that people complain about it. Certainly it is easier to complain about noise than to formulate complaints about abstractions, such as the presentation of information on displays, even though the latter may be far more important to the task. The author once visited an air-traffic-control tower to make a survey of ergonomic problems. The operators first complaints were of 'uncomfortable chairs'. Later we found severe problems with the information that was presented, for example the design of displays that illustrated how aeroplanes were taxiing and lining up on the ground for take-off. The modification of the information displays was by far the most important ergonomic problem. But the issue is somewhat abstract, difficult to think of, and difficult to talk about.

14.6.1. Interference of Noise with Spoken Communication

Noise is a well-qualified problem because it disrupts communication, and some ergonomics standards have postulated that the noise level should be no greater than 55 dBA in office environments, in order to facilitate communication (Human Factors Society, 1988).

There are two common methods for evaluating the effect of noise on communication: preferred noise criteria (PNC) curves, and preferred speech interference level (PSIL).

14.6.1.1 Preferred Noise Criteria (PNC) Curves

This methodology was developed by Beranek *et al.* (1971) (Figure 14.6). The curves are based on office workers' subjective ratings of noise. The ratings were given during several experiments done to investigate which frequencies in the noise are particularly disturbing to speech communication. The PNC curves in Figure 14.6 represent equal-sensitivity (isosensitive) curves to noise of different frequencies. For the higher frequencies the curves come down. The human sensitivity to high frequency noise is greater and less sound pressure is needed to perceive the noise, but for the lower frequencies the human ear is less sensitive and much greater sound pressure is needed to perceive the noise.

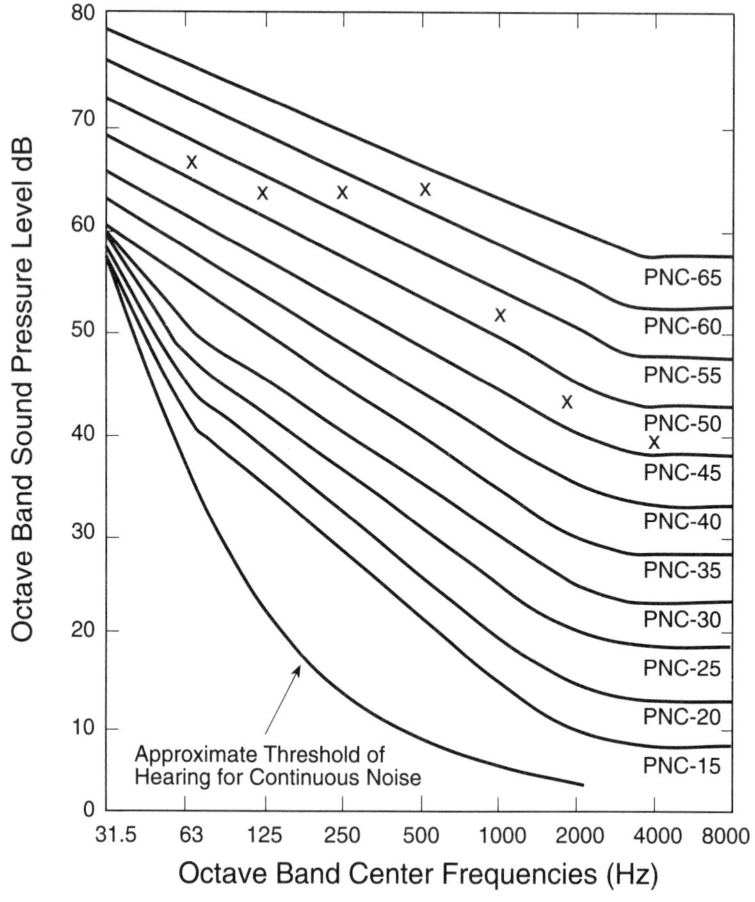

Figure 14.6 Preferred noise criteria (PNC) curves are isosensitive curves for noise of different frequencies (the numbers after PNC is the sound pressure in decibels at 1000 Hz). To evaluate a noisy environment, octave-band readings are obtained and plotted on top of the PNC curves

Table 14.4 Recommended PNC curves and sound pressure levels for different listening conditions

Acoustical requirements	PNC	Approximate dBA
Excellent listening conditions	≤20	≤30
Good listening conditions	20–35	30–42
Moderately good listening conditions	35–45	42–52
Fair listening conditions	40–50	47–56
Just acceptable speech and telephone communication	50–60	56–66

The PNC curves are used to evaluate the acoustical requirements for different tasks. Some recommended PNC curves and the approximate sound pressure levels are presented in Table 14.4.

At higher PNC values (around 50 and 60) it becomes very difficult to communicate with other individuals. The PNC values are different from dBA, because dBA is an average weighting across the entire sound spectrum, whereas the PNC provides an evaluation throughout the noise spectrum. As an example of using PNC curves, assume that in speech communication a sound-level meter is used to obtain octave-band readings of the noise. Assume further that we have selected PNC = 60 as a criterion for evaluation. The criterion is exceeded for 500 Hz, but otherwise the noise level is acceptable (see Figure 14.6).

14.6.1.2 Preferred Speech Interference Level (PSIL)

The PSIL is the most common method for rating the speech interference effects of noise (Webster, 1969). The PSIL value is first calculated by averaging the sound pressure levels (in decibels) of octave bands centred on 500, 1000 and 2000 Hz. Thus, if the levels of noise were 65, 70 and 75 dB, respectively, the PSIL would be 70 dB. The PSIL can give a fairly good approximation of the impact of noise having a flat spectrum. However, if there are irregularities in the spectrum, it loses some of its use because the simple average of the three octave bands cannot characterize the noise.

The PSIL value is evaluated using a graph. In Figure 14.7 the distance from a speaker to a listener is given as a function of the PSIL value. The necessary speech level is then characterized as: normal, raised voice, very loud voice, shout, maximum vocal effort, or limit for amplified speech. Thus, for example, if the PSIL value is 65 dB and the distance to a listener is 8 ft the speaker would have to talk with a very loud voice. PSIL has also been used to characterize office

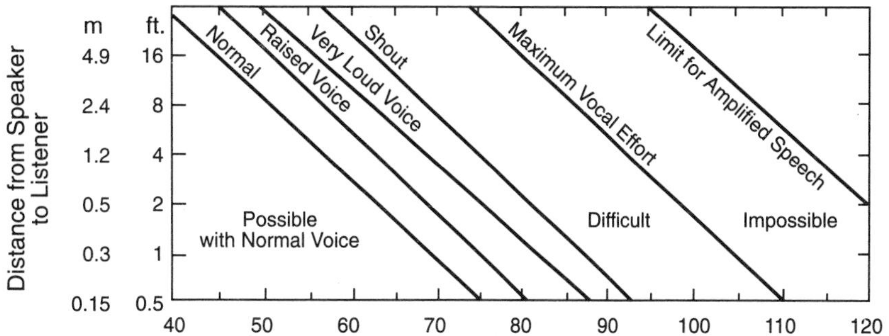

Figure 14.7 Voice level and the distance between the speaker and the listener as a function of PSIL noise level

communication in private offices and secretarial offices (Beranek and Newman, 1950). Of particular interest is the effect of noise on telephone use. For a PSIL value greater than 60 dB it is difficult to use a telephone, and for a value greater than 76 dB it is impossible to talk on the telephone.

Exercise: How to Use PSIL. To evaluate the ease of communication in an industrial plant, the noise was measured for three octave-bands: 500, 1000 and 2000 Hz. The recorded noise levels were 75, 80 and 82 dBA, respectively.

1. Calculate the PSIL value!
2. Using the values in Figure 14.7, what is the maximum distance at which two individuals can communicate without raising their voices.
3. Given the social unacceptability of a very close distance, what would be the necessary speech level if the distance was 1.0 m?

Chapter 15

Shift Work

Shift work is not a new phenomenon. Scherrer (1981) reported that in Ancient Rome, transportation of goods had to be performed at night in order to reduce traffic congestion. However, it is only during the last century, after Edison's invention of the lamp, that shift work has become widely adopted in industry. This is concomitant with several trends in industry and society:

1. *Process industries*. Many modern industries such as power plants and steel works cannot close at night.
2. *Economic pressures*. Companies often prefer to introduce a second and a third shift because production machinery is expensive and cannot be duplicated. In addition, shift work makes it possible for individuals to work overtime, which is less expensive and is often perceived as less risky than recruiting additional employees.
3. *Service sector demands*. In the service sector there are many types of job where people are needed around the clock (nurses, physicians, policemen, transportation workers, and restaurant employees).

In this chapter we take a broad definition of shift work as being anything outside the hours of 7.00 am and 6.00 pm (Monk and Folkard, 1992). With this broad definition, approximately 20–30% of the workforce participates in shift work. A later study found that 22% of the working population were shift workers: 16% of full-timers and 47% of part-timers (Mellor, 1986). Similar figures have been estimated in the UK and Sweden (Monk and Folkard, 1992). In the USA, Tasto and Colligan (1977) estimated the percentage of shift workers in several job categories (Table 15.1).

Table 15.1 Percentage of shift workers in various industries in the USA

Type of industry	Shift workers (%)
Postal workers	45.8
Food workers	42.7
Hospitals	36.9
Rubber/plastic	35.0
Railroad	32.7
Tobacco	32.8
Printing	28.5
Welfare	21.8
Chemical	19.7
Education	17.0

There are two types of operation: around-the-clock, usually involving three shifts; and operations involving fewer hours. The around-the-clock operation, and in particular the hours from 12 midnight to 4.00 am cause severe problems in terms of health, fatigue, and lost productivity, and these problems are the major focus of this chapter.

Shifts are usually designated as 'morning shift', 'afternoon shift' and 'night shift'. There are other common names: 'day shift', 'swing shift', and 'graveyard shift', or simply 'shift 1', 'shift 2' and 'shift 3'. In this chapter we use the first designation, as illustrated in Table 15.2.

15.1 Example: How Not to Schedule Shift Work

In 1981, the author visited an underground metal mine in the southern USA. During interviews with the workers it was obvious that many of them suffered fatigue from participating in shift work. It turned out that there were only two shifts, and the working hours had a beautiful symmetry:

Shift 1: 7.00 am–3.00 pm.
Shift 2: 7.00 pm–3.00 am.

We asked a manager why there were 4 hours of non-work starting at 3.00 pm. He gave the following explanation: the work procedures were identical for both crews. At first when they arrived they would transport ore and rocks which had just been blasted by the previous shift. Then they would start drilling holes for blasting, and at the end of the shift they would blast. Many years ago it used to be that blasting agents produced lots of smoke, and it was necessary to ventilate the mine for 4 hours before the next crew could come in. However, with modern types of blasting agent ventilation is no longer necessary. We told the manager that the problems with shift work would be eliminated if the second crew could work from 3.00 pm to 11.00 pm. But they did not even want to try: 'We would hate to renegotiate the contract with the union!'

15.2 Circadian Rhythms

The basic physiological problem with shift work is that the body establishes a 24-hour rhythm which is difficult to change. Figure 15.1 illustrates the so-called 'diurnal' or 'circadian' changes in oral temperature over 24 hours. The temperature is at a maximum at about 4.00 pm and at a minimum at about 4.00 am. Many other body mechanisms (heart rate, breathing rate, body temperature, excretion of many types of hormones, and urine production), follow the same sinusoidal pattern (Chapanis, 1971). Assume that a person starts working the night shift (10.00 pm–6.00 am) instead of the morning shift (6.00 am–2.00 pm). It would then take about 1 week to flatten out the sinusoidal curve and about 3 weeks to reverse the waveform. However, as illustrated in Figure 15.1 the pattern is never quite reversed. The circadian changes are smaller for a person who works the night shift as compared with those for a person who works the morning shift.

Table 15.2 Typical working hours in shift work

Name of shift	Typical working time	
Morning shift	6 am–2 pm	(6.00–14.00 h)
Afternoon shift	2 pm–10 pm	(14.00–22.00 h)
Night shift	10 pm–6 am	(22.00–06.00 h)

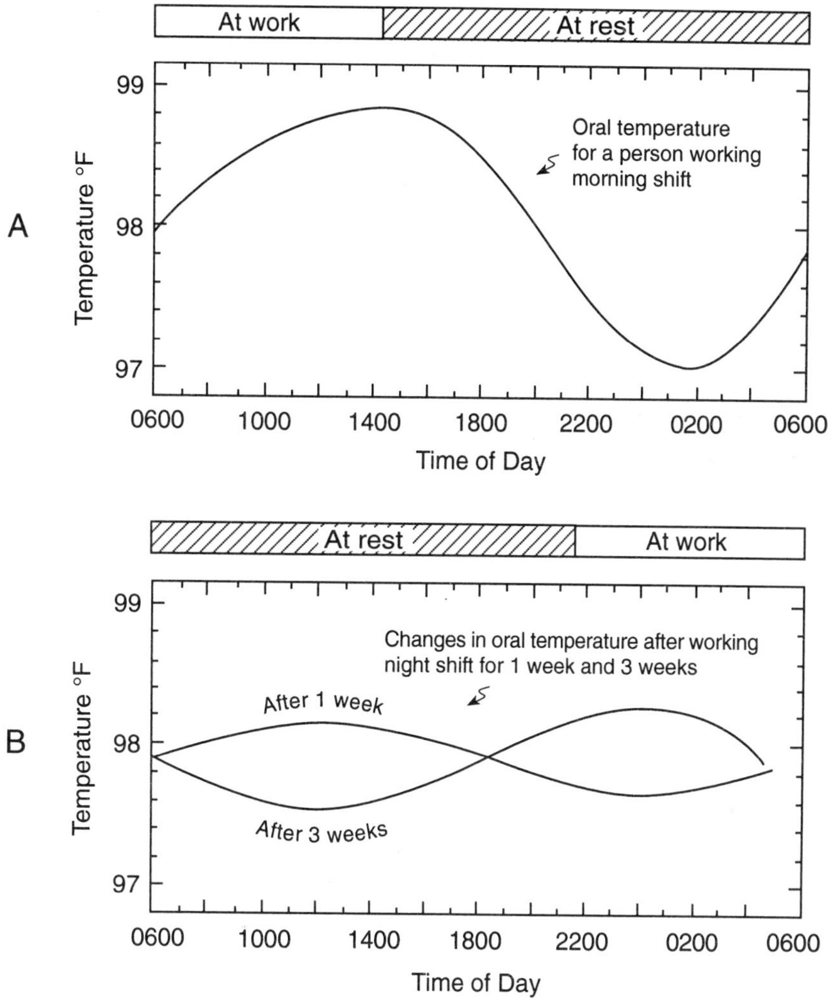

Figure 15.1 Diurnal (or circadian) rhythm of oral temperature. (A) The normal pattern for day work. (B) There is a flattening after one week and reversal of the curve after 3 weeks of working on the night shift

In other words, it seems to be impossible to adjust totally to night-time work.

There are many reasons for this lack of adjustment. The most important component may be daylight. Daylight is a very forceful cue in indicating the time of day. In German, this phenomenon is referred to as a *Zeitgeber* (literally, time-giver). Some recent research has shown that exposure to daylight levels (more than 2000 lux) of illumination increases alertness during night shifts, and suppresses the production of melatonin (a sleep-inducing hormone). But there are also many environmental and social *Zeitgebers* (Monk and Folkard, 1992). It is easier to sleep during the night-time because there is less disturbing noise and there are no social activities. On the other hand, a night worker suffers more from daytime noise and daytime activities, and family and friends also disturb the sleeping pattern.

15.3 Problems with Shift Work

There are many problems with working the night shift. Some of these problems have been well documented, whereas others have been suggested but not yet verified by research (Table 15.3). Some of the items listed in the Table 15.3 warrant some comment. It is evident

Table 15.3 *Typical problems associated with working the night shift*

Fatigue. On average a night-shift worker sleeps 1.5 h less
Health disorders. Stomach problems, digestive disorders, and possibly an increased rate of cardiovascular disease
Disruption of social life. With family, friends, labour unions, meetings, and other gatherings
Decreased productivity. More for knowledge-based tasks than skill- and rule-based tasks
Safety. Accident rates may increase

that shift workers have a much higher rate of stomach problems than daytime workers (Monk and Folkard, 1992). Part of the problem is that the sensation of appetite is tied to the circadian cycle. Shift workers are hungry at the 'wrong times', and go to the toilet at the 'wrong times'. The appetite is suppressed while people are asleep and is greater during daytime. Individuals starting on the night shift will carry their daytime habits along until they have adjusted. In addition, shift workers eat more junk food than do daytime workers. One reason for this may be that the company cafeteria is closed and there are no cooked meals available. A shift of the circadian cycle also disturbs the digestive functions.

One of the basic problems with research on shift work is that, while some individuals like it, about 20% of the population has severe difficulties and will never adjust. Perhaps their constitution is not robust enough to cope with shift work, and perhaps those who remain in shift work are physically stronger and have better health. Therefore the population of study may be biased to start with, and there cannot be a fair scientific comparison.

The possibility of an increased rate of cardiovascular disease in shift workers has been suggested, but this is difficult to verify. The most convincing study of increased heart disease was performed in Sweden (Knutsson *et al.*, 1986). This study involved 50 workers in a paper mill. It showed that after 10–15 years of exposure to shift work, the risk of heart disease was doubled compared with a population of workers on day shift. But there were many uncontrolled factors. In addition to shift work there might have been differences related to life-style, diet, and so forth, although some of these factors were taken into account in the study. To draw firm conclusions more research is necessary.

The disruption of social life is another important consequence of shift work. Night work can make it impossible to participate in gatherings of family and friends and other social functions. This is one of the major reasons why several countries in Europe propose a fast rotating shift-work schedule, with 2 or 3 days at most on each type of shift (Table 15.4).

15.4 Effects on Performance and Productivity

It has been difficult to establish in research whether productivity is reduced during the night shift. One of the problems is that the type of work tasks are often different, so there cannot be fair comparisons with daytime work. For example, some plants schedule maintenance work during the night shift, whereas in other plants maintenance work

Table 15.4 The effect of shift work on social activities and leisure activities (adapted from Knauth et al., *1983)*

Not enough time for	Shift workers (%)	Day workers (%)
Social events	87	22
Cultural events	72	11
Friends	80	13
Family	72	11
Hobbies	67	17

is performed during the day shift. There is also a lack of supervisors and managers during the night shift, which means that group morale can suffer.

The consensus from the research is that simple skill-based and rule-based tasks do not suffer as much during shift work. Cognitive, knowledge-based tasks requiring 'deep thinking' are more affected. A study by Bjerner and Swensson (1953) evaluated records of error frequency in reading meters at a gas company. The error frequency was greatest at 3.00 am. Browne (1949) evaluated the speed of switchboard operators. The slowest responses were obtained between 3.00 am and 7.00 am.

Several studies have pointed to the effect of circadian rhythms on accidents. Folkard *et al.* (1978) showed that the frequency of minor accidents is the greatest at 5.00 am. Harris and Mackie (1977) investigated accidents involving US interstate truck drivers, due to falling asleep. They found that the accident rate was 20 times as high at 5.00 am as at 12 noon.

One of the most quoted, although least conclusive, events is that the Three Mile Island nuclear accident occurred during the night shift. The occurrence of this event was traced to human error and may not have occurred during daytime (Monk and Folkard, 1992).

15.5 Improving Shift Work

Guidelines that can be used for scheduling shift work and for selecting individuals to participate in shift work are listed below.

15.5.1 Type of Work

The length of the shift should be related to the type of work. For light work a 12-hour shift could be contemplated. In fact, most workers like 12-hour shifts (Miller, 1992). There is better job satisfaction, improved morale, and reduced absenteeism. But alertness and thus safety may decline, and workers may work at a slower pace.

For heavy physical or complex mental (knowledge-based) work shifts should be no more than 8 hours, and may be only 6 or 7 hours, during the night.

Visual inspection and visual monitoring is extremely difficult during the night time. This is a low vigilance task. The arousal level is low even during daylight hours and at night time many operators simply fall asleep. Rohmert and Luczak (1978) investigated operators sorting letters in the German Post Office. After working for only 2 hours on a night shift the fatigue became overwhelming. In addition, during the critical hours of 3.00–5.00 am the error rate in sorting letters increased significantly. Due to the problems with fatigue and because missorted letters are extremely costly for the postal system, it was decided to abolish the night shift – a radical solution for any operation. For these reasons visual inspection and quality control should not be scheduled for the early morning hours.

Miller (1992) suggested that the number of hours could be reduced for the night shift. It might be advisable to use a shift schedule of 8-hour morning, 9-hour afternoon, and 7-hour night (8M–9A–7N) or, alternatively, 8M–10A–6N or 9M–9A–6N. This may allow the worker to deal more appropriately with the greater amount of stress experienced during the night shift.

15.5.2 Shift Work Schedules

There is an infinite number of ways of arranging a shift-work schedule. Here we restrict ourselves to the most difficult case: 7 days of operation using four shift crews. Knauth *et al.* (1979) pointed out that the 40-hour working week is cumbersome and limiting, and that a 42-hour week allows an even distribution of worktime across workers on all shifts, because:

$$7 \text{ days/week} \times 24 \text{ hours/day} = 168 \text{ hours/week}$$

$$42 \text{ hours/crew} \times 4 \text{ crews} = 168 \text{ hours/week}$$

Thus, the week is nicely divided into four 42-hour segments.

In the German and Scandinavian countries there is a clear preference for fast, forward-rotating shift schedules. The philosophy is that the number of consecutive night shifts should be as few as possible. Preferably, there should be only one consecutive night shift in a shift schedule. In the schedule in Table 5.5 several important principles have been incorporated:

- It takes 4 weeks to go through the cycle. The shorter the cycle, the easier it is for the worker to keep track of it.
- After each night shift there is at least 24 hours of rest.
- The long weekend at the end of the first week is much appreciated.
- The shift assignments rotate forward: from morning to afternoon to night.

Forward rotation is advantageous because the true diurnal cycle is closer to 25 than 24 hours. That is, people have a tendency of wanting to go to bed 1 hour later every night. This has been proven in investigations where people live in isolation for a long period of time without any time cues (as if they are living in a dark, isolated cave).

The main philosophy behind this shift-work pattern is that workers are supposed to remain adjusted to the daytime schedule. Usually it is possible to work a single night shift without being overly tired. Of course the one disadvantage of this shift-work pattern is the

Table 15.5 A rapid forward-rotating 8-hour shift system with four crews and a 4-week cycle for a 42-hour week: each crew will work 21 shifts of 8 hours each (total 168 hours)

Week	Mon.	Tues.	Wed.	Thur.	Fri.	Sat.	Sun.
1	N	–	M	A	N	–	–
2	–	M	A	N	–	M	M
3	M	A	N	–	M	A	A
4	A	N	–	M	A	N	N

M, Morning shift; A, afternoon shift; N, night shift; –, rest.

sequence of three nights at the end of week 4 and beginning of week 1.

Labour unions in Germany and the Scandinavian countries have claimed that this type of schedule improves family life and social life (Rutenfranz *et al.*, 1985). But the tradition elsewhere in the world is different. In the USA it is common to have a slowly rotating shift schedule with one week devoted to each shift. Monk and Folkard (1992) suggested that this might be the worst possible policy, since there is insufficient time for the body to adjust to the new work patterns (see Figure 15.1). A much slower speed of rotation with 3 weeks or more in one shift would allow circadian adjustment. The main controversy has been discussed in detail by Monk (1986) and revolves around the loss of nocturnal orientation during free weekends, which break down the adjustment to the night-time schedule.

Two alternative fast-rotating shift-work schedules, the so-called 'metropolitan rota', or '2–2–2 shift system', and the 'continental rota', or '2–2–3 shift system', are displayed in Table 15.6 (Knauth *et al.*, 1979). The numbers refer to the number of days on each shift. We provide these examples to illustrate the endless number of combinations that exist for shift-work schedules. However, in the European tradition, the schedules illustrated in Tables 15.5 and 15.6 are among the better ones.

There are social advantages in starting the morning shift either at 7.00 am or 8.00 am instead of 6.00 am; the family can have breakfast together. The preferred starting hours would then be 7–15–23 or 8–16–24.

15.5.3 Selecting Individuals for Shift Work

Some individuals, although they volunteer to participate in shift work, may eventually have difficulties in coping. Usually they are at a disadvantage from the very beginning. There are several factors which can be used to predict if individuals can be expected to have difficulties with shift work (Tepas and Monk, 1986). Managers and workers should be informed about these factors, since they are linked to satisfaction and success on the job (Table 15.7).

Table 15.6 The metropolitan rota (2–2–2) and the continental rota (2–2–3) shift systems: both systems assume 4 crews and 42-hour weeks; The metropolitan rota has an 8-week cycle and the continental rota a 4-week cycle

Metropolitan rota Week No.	Mon.	Tues.	Wed.	Thur.	Fri.	Sat.	Sun.
1	M	M	A	A	N	N	–
2	–	M	M	A	A	N	N
3	–	–	M	M	A	A	N
4	N	–	–	M	M	A	A
5	A	N	N	–	–	M	M
6	A	N	N	–	–	M	M
7	A	A	N	N	–	–	M
8	M	A	A	N	N	–	–
Continental rota							
1	M	M	A	A	N	N	N
2	–	–	M	M	A	A	A
3	N	N	–	–	M	M	M
4	A	A	N	N	–	–	–

M, Morning shift; A, afternoon shift; N, night shift; –, rest.

Table 15.7 Individual factors that are likely to cause problems in adapting to shift work

- People living alone do not adjust as easily
- More difficult for people with gastric or digestive disorders
- People with inadequate sleeping facilities suffer more
- Over 50 years of age
- Morning-type individuals (larks)
- Second job or heavy domestic duties
- Epileptics

Family members usually support a shift worker and make concessions. A wife of a shift worker had bought a white-noise generator for her husband to diminish the impact of noise during the daytime; the bedroom had special curtains to make it completely dark; and meals were served at special times to help her husband to adjust to the shift-work schedule.

With increasing age it seems that individuals become more set in their circadian rhythm. There is also a change towards a pattern of 'morningness', indicating that individuals tend to go to bed earlier and wake up earlier. Morningness is indeed one of the greater obstacles to shift work. Öquist (1970) established differences between 'morning types' and 'evening types', and Horne and Östberg (1976) have published a questionnaire which can be used to distinguish between morning and evening types. This questionnaire can be used to help select 'evening types' who are more suitable for shift work.

Several medical conditions could disqualify an individual from shift work. People with gastrointestinal problems get worse. Epileptics have a higher rate of seizures during the night shift.

15.6 Recommended Reading

An excellent text on shift work is: *Making Shiftwork Tolerable* (Monk and Folkard, 1992).

Chapter 16

Whole Body Vibration

In today's occupational environment, mechanical vibration is frequently encountered. It can present a health hazard. There are two major kinds of vibration: whole body vibration and hand vibration. The latter is commonly referred to as 'segmental vibration', implying vibration of the extremities. In addition to these there is a third phenomenon - sea sickness, which involves exposure to slow vibrations in the range 0–1 Hz.

A common source of whole body vibration is vehicles of all types including forklift trucks, long-haul trucks, earth-moving equipment, and other industrial moving machines.

Hand vibration or segmental vibration is often induced by hand-held tools such as power drills, saws, jack hammers, concrete vibrators, and chain saws. These are dealt with in the Chapter 8. In this chapter we give an overview of the most common problems related to vibration.

16.1 *Sources of Vibration Discomfort*

A common source of whole body vibration is from transportation vehicles where drivers are exposed to a vibration generated by the vehicle and the roadway. Figure 16.1 shows a seated driver, illustrating that different parts of the body have different resonant frequencies. For the shoulder and the stomach the resonant frequency is 3–5 Hz. This may be the reason why this particular frequency range produces the greatest reported discomfort.

Laboratory studies have confirmed that vibrations between 3 and 5 Hz are likely to be physically uncomfortable at an acceleration level

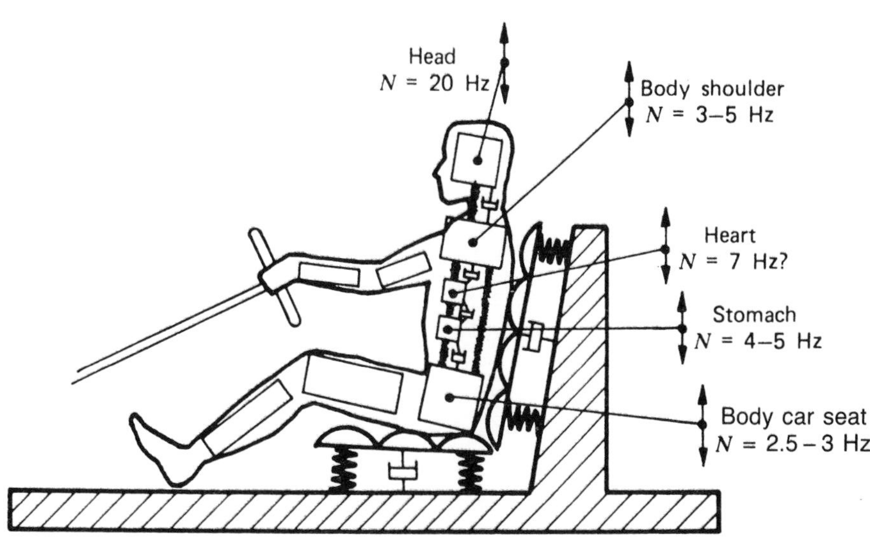

Figure 16.1 The resonant frequencies of different parts of the body of a seated driver

of approximately 0.1*g*, painful and distressing at intensities of about 1*g*, and will cause injuries if the acceleration exceeds 2*g*. These types of finding form the main background for the present ISO (International Standards Organization) standards for vibration (Figure 16.2) (Mackie *et al*., 1974; Gruber, 1976).

Hansson *et al*. (1976) studied exposure to whole-body vibrations of drivers of 44 industrial trucks. He found that using ISO Standard 2631 for exposure limits (ISO, 1976) six of the industrial trucks presented a risk to health if exposure lasted for 8 hours. Vibration was fatiguing and reduced the work capacity of the drivers in two-thirds of the trucks studied (according to ISO standards).

Large and heavy trucks exposed the drivers to lower frequencies than did small and light trucks. Obviously, the vibration characteristics of similar machines vary considerably depending upon the design. Hansson concluded that manufacturers and designers of trucks are not always well informed about the implications of different design alternatives on whole body vibration.

It is not only the amount of physical energy, there are also several *psychological factors* that can greatly influence the discomfort of vibrations, including (Mackie *et al*., 1974):

1. *The nature of the task*. For example, riding in a recreational boat is usually associated with pleasure, although the same magnitude vibrations would be perceived as very stressful in an industrial environment.
2. *The person's degree of training or familiarity with the task*. For example, a skilled horseback rider can compensate for much of

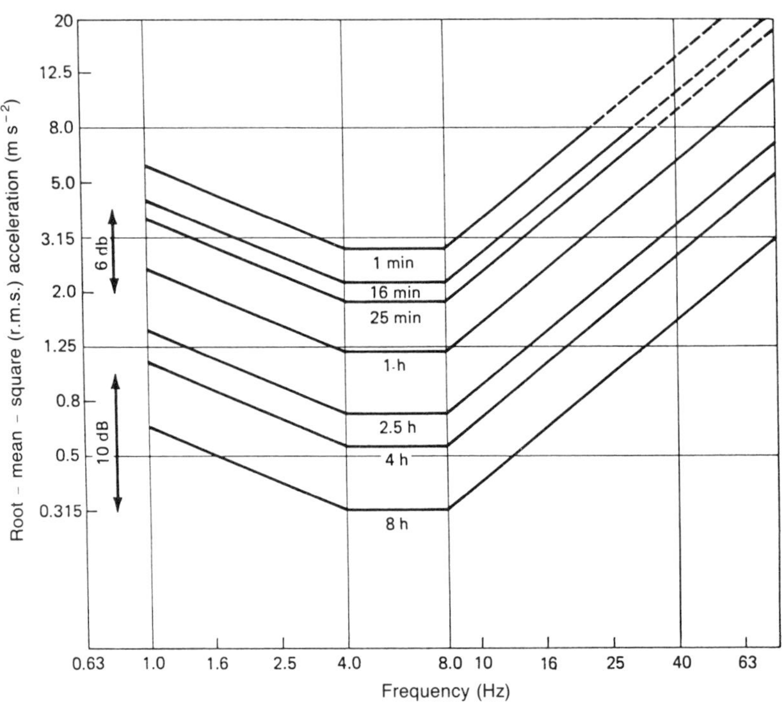

Figure 16.2 ISO Standard 2631 prescribes exposure limits of vibration for 8 hours of work and less than 8 hours. The figure illustrates exposure limits for vertical (y direction) vibration. There are similar regulations in the x *and* z *directions. To obtain exposure limits for reduced comfort subtract 10 dB: For exposure limits to avoid tissue damage add 6 dB*

the vibrations by rhythmically contracting certain muscles. Likewise, an industrial worker can compensate for some of the movement of a vibrating forklift truck or piece of industrial machinery. There are also individual differences in sensitivity to vibration. Particularly important is that heavy individuals suffer more from vibration than do light individuals.

3. *The presence of other stressors acting in combination*. For example, vibration in combination with noise produces a greater level of stress than vibration alone or noise alone (Poulton, 1979). This will effect the physiological arousal of the individual, which in turn has implications for the performance level (Figure 16.3).

In addition to the discomfort effects of vibration, there are several reputed health effects such as various spinal, anal-rectal, and gastrointestinal disorders (Fothergill and Griffin, 1977). However, these have been difficult to verify in research. Most of the evidence comes from epidemiological investigations of truck drivers and heavy-equipment operators (Seidl and Heide, 1986). A large US study of truck drivers reported that drivers complain about these problems, but there are other possible factors that could also contribute, such as extended sitting and poor eating habits.

Exposure to vibration also induces *physiological responses*. The most basic physiological response to a moderate level of vibration is an increase in heart rate; about 10–15 beats/min above the resting level. The heart rate returns to its normal level after the vibration ceases. Blood pressure can also increase, particularly for vibration frequencies around 5 Hz. Some investigations have revealed a slightly increased breathing rate, and oxygen consumption. These changes may be related to increased muscular activity which is induced by vibration. As mentioned above, people exposed to vibration will often contract muscles to compensate.

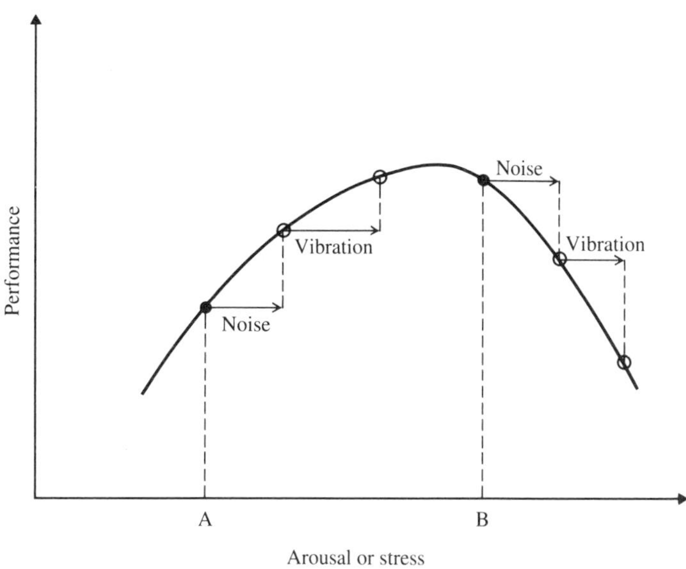

Figure 16.3 Yerkes–Dodson's law can be used to illustrate the additive effect of two stressors on performance. The initial arousal level (A or B) is crucial to performance

One of the most notable findings is that at vibrations of about 10–25 Hz the visual acuity level decreases. This frequency range is thought to represent the resonant frequency of the eyes, and as a result there is often a reduction in the operator's performance level (Grether, 1971; Collins, 1973).

Whole body vibration also effects the motor performance and muscle control. Hornick (1973) reviewed the literature and concluded that for tracking experiments (with a joystick) the tracking errors could increase by 40% in a vibrating environment. By supplying an arm support the errors were reduced to about half.

Chapter 17

Design for Manufacturing Assembly

Ergonomics professionals working for manufacturing companies have in the past specialized in two areas: design of industrial workstations and design of products to improve functionality and usability. In this chapter we propose a new field of activity: the study of the effects of product design on the types of job created in the assembly of the product. The basic concept is that, through product design, jobs are created in manufacturing. It is then important to design products so that they are easy to assemble and do not create safety hazards. One must also distribute the manufacturing tasks between manual labour and automated processes. This distribution of tasks or 'task allocation' must be productive for the company, and it must create satisfying jobs for the employees.

17.1 The Desire to Automate

During the 1980s, manufacturing engineers vigorously pursued opportunities to automate, and sometimes the results were very disappointing. During this time, General Motors invested $80 billion in automated manufacturing, but at least 20% of their spending on new technology failed (*The Economist*, 10 August, 1991). Other major companies had similar experiences. In many cases the surprising reason was that manual labour, with its greater flexibility and adaptability, can outperform automation. The focus on automation has not been serving our interests. Automation does not by itself increase productivity and job satisfaction. Unfortunately, automation is what engineers take an interest in. Modelling of human work is difficult, and principles of allocating work functions between humans, machines and computers are not well understood by the engineering community.

Robots were first used in industry for fairly simple tasks such as welding and painting. At the beginning of the 1980s there was increasing interest in using robots for assembly. Early on there was a realization that robots can only be used for fairly simple assembly tasks which are easy to describe and program. In order to enhance the utility of robots, assembly had to be simplified, and it became necessary to redesign products so that they were easy to assemble by automation. In the last 10 years many design guidelines have been published, which prescribe the design of parts that are easy for a robot to assemble. This type of product design is referred to as 'design for automation' (DFA) or 'design for manufacturability' (DFM) (Boothroyd and Dewhurst, 1983). However, sometimes the redesign of a product leads to a very surprising outcome, as we illustrate in the following example.

Example: The Assembly of a Paper Picking Mechanism
In the early 1980s, IBM Corporation was manufacturing copy machines at the plant in Boca Raton, Florida. The paper picking mechanism

in the copy machine had 27 parts (Helander and Domas, 1986) (Figure 17.1).

One problem was that individual parts require individual feeding mechanisms, which standardize their presentation so that they are easy for the robot to grip. Twenty-seven parts were simply too many, because the entire work envelope would be filled with part-feeding mechanisms. After the redesign the paper picking mechanism had 14 parts, 13 of which could be assembled by robots and automation. The fourteenth part required a complex insertion motion and had to be put in place using manual labour (thereby creating a highly repetitive task). The surprising outcome of the redesign was that manual assembly of the mechanism became so simple, that the cost of using robots could no longer be justified. This product is still assembled manually.

From our perspective it is ironic that only the introduction of automation has compelled engineers to investigate the principles of manual assembly. Throughout the history of manufacturing, engineers have taken for granted that workers can adapt to any situation. Engineers have ignored opportunities for ergonomic improvements, which could increase productivity as well as operator comfort. Only in the last 10 years, through the advancement of automation have

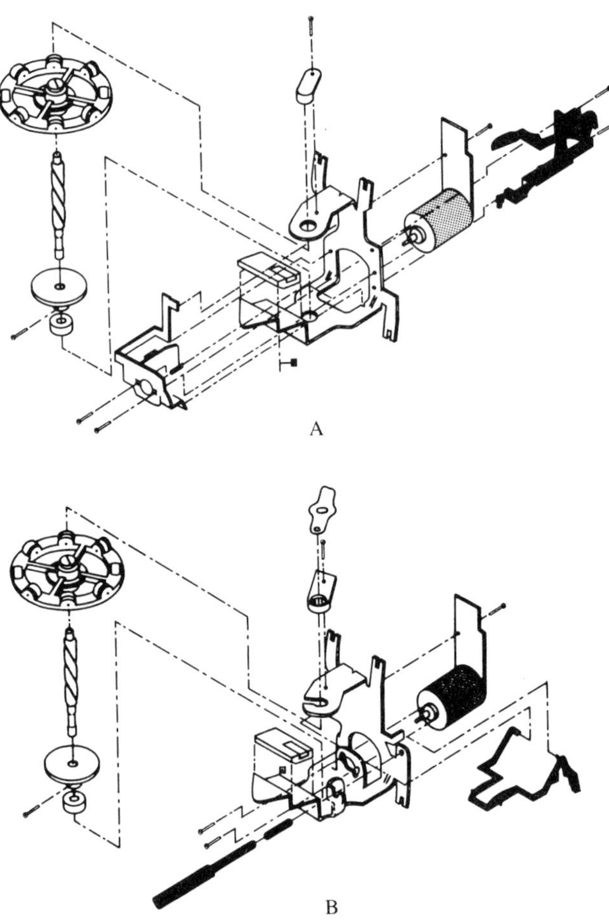

Figure 17.1 (A) The origin design of the paper picking mechanism had 27 parts. (B) The redesigned mechanism has 14 parts

engineers been forced to consider alternative production methods – in this case manual labour.

One may question under what circumstances automated devices are actually more productive than human labour. Within the IBM Corporation there are several similar cases of product redesign, where manual labour has ultimately proven more cost-effective. One case study is the printer manufactured at IBM in Charlotte, North Carolina, which is mentioned Chapter 1 (see also: Genaidy *et al.*, 1990; Mital, 1991). In these cases manual assembly was faster and the introduction of automation could not be justified. Could it be that design for human assembly (DHA) is a viable method, and we do not need robots? As usual in ergonomics, it depends on the task.

In the rest of this chapter we provide an overview of guidelines that may be used in product design in order to simplify both automated assembly and manual assembly. The information is collected under four headings:

(1) What to do and what to avoid in product design.
(2) Boothroyd's method for the redesign of products.
(3) Use of predetermined time systems to diagnose product design.
(4) Human factors design principles applied to product design.

17.2 What to Do and What to Avoid in Product Design

In this section we provide examples of product design features that simplify assembly. Many of them are used for automation and have been published in guidelines for DFA. They apply equally well to manual assembly (Helander and Nagamachi, 1992).

17.2.1 Using a Base Part as the Product Foundation and Fixture

Design the product with a base part as the foundation and fixture for other parts. It should be possible to assemble the other parts from one direction, preferably from above (Figure 17.2). It is also advantageous to use fasteners which are inserted from one direction, either from the front or from above. The base part should also serve as a fixture. If this arrangement is not feasible, pins can be used so that the base part can be easily positioned on a fixture as in Figure 17.2. If this is not possible, a specially designed fixture is used.

To make the product easy to transport, it should have a flat bottom and a simple shape.

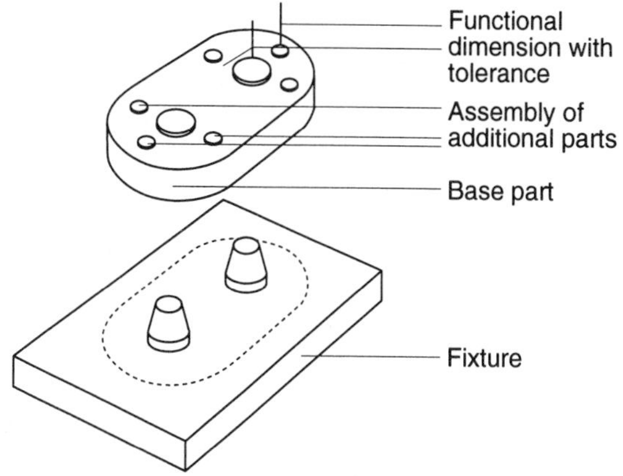

Figure 17.2 Provide a simple and reliable fixture for the base part. If possible the base part should also serve as a fixture

17.2.2 Minimizing the Number of Components and Parts

1. Integrate or combine parts, since they take less time to organize and less time to assemble. In some cases an entire subassembly can be replaced by a single part (compare with modular design in electronics). Integrated parts may be complex to handle, but they reduce the number of operations (Figure 17.3).

   Holdbrook and Sackett (1988) noted that it is *difficult* to combine parts if:

   - Parts move relative to each other.
   - Parts are required to be of different materials.
   - Parts must be separate for maintenance and sevice reasons.
   - Parts are necessary to enable the assembly of remaining parts.

   Combined parts can often be fabricated using plastic injection moulding. Another advantage with plastic parts is that they can easily be provided with chamfers, notches, and guides which are helpful in assembly. Metal parts can also be moulded or mounted into plastic parts. The elastic property of thermoplasts (e.g. nylon) can be used to form snap joints, integral springs and integral hinges. Thermoplasts can also be used to straighten other parts and to eliminate clearances.

2. Eliminate or minimize different types and sizes of fasteners:

   - Use snap and insert assembly. If possible, design integral fasteners and clips into parts so that no screws are required, as in Figure 17.4.
   - Minimize the various types and sizes of screws (Figure 17.5). Fewer number of parts decrease the number of part bins, which saves space. A smaller number of bins will also decrease the operator's choice–reaction time between bins. In addition, fewer

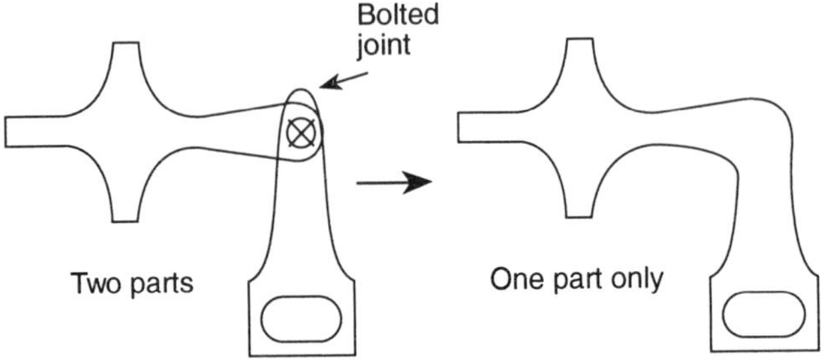

Figure 17.3 Integrate or combine parts

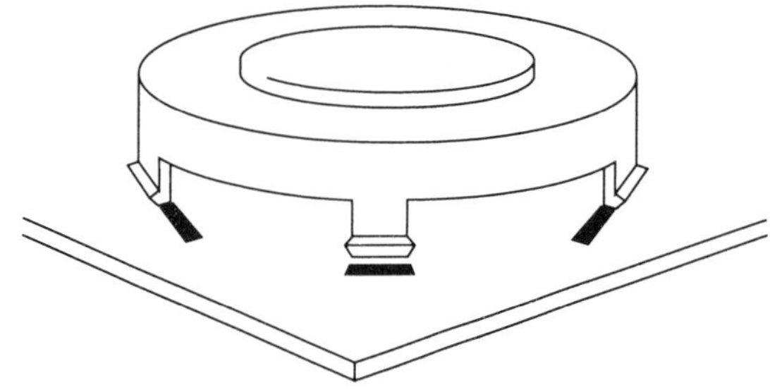

Figure 17.4 Use snap and insert assembly

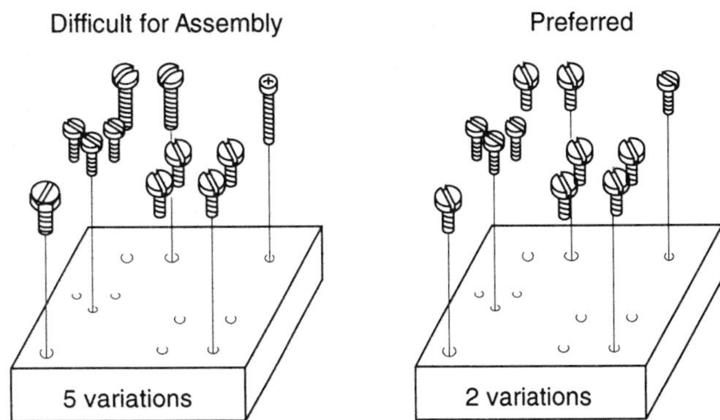

Figure 17.5 Minimize the various types and sizes of screws

parts will reduce the number of hand tools, which in turn decreases handling time and space requirements.

3. Do not use small parts such as washers. This requirement which is mandatory for robotic assembly, also simplifies manual assembly (Figure 17.6). The use of washers increases the manual handling time. Their use may also make it necessary for the operator to use pinch grips, which have been implicated as a cause of cumulative trauma disorder.

17.2.3 Facilitating Handling of Parts

1. Improve parts handling by using parts that are easy to grip (Figure 17.7).
2. Avoid using flexible parts, such as wires, cables and belts, because they are difficult to handle. Sometimes components can be plugged together in order to eliminate the use of connecting wires.
3. Avoid parts which nest or tangle. Close open ends and make part dimensions large enough to prevent tangling. For example, use springs with closed ends rather than open ends (Figure 17.8).

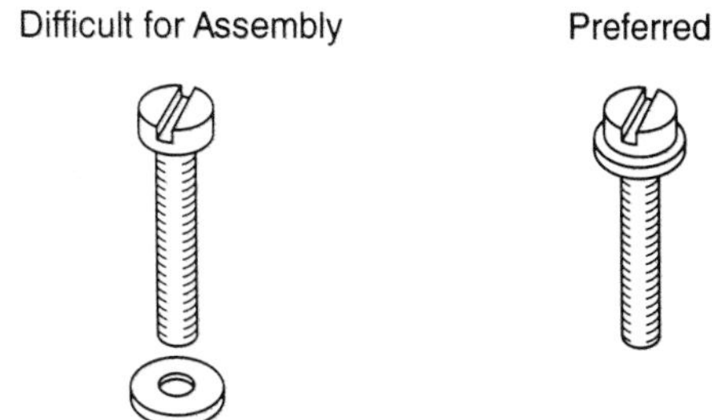

Figure 17.6 Do not use small parts that are difficult to handle, such as washers

Figure 17.7 Improve parts handling by making parts easy to grip

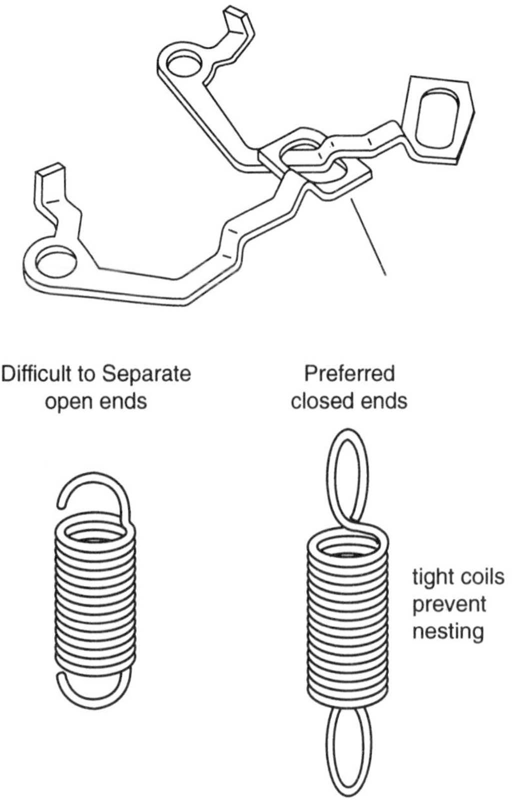

Figure 17.8 Avoid parts that nest or tangle

17.2.4 Facilitating Orientation of Parts

1. Use symmetrical parts, because they are easy to orient (Figure 17.9). The use of symmetrical parts reduces information processing, since the operator does not have to decide whether to turn the part round. It also reduces manual handling time.
2. If asymmetric parts are used, provide visual aids for orienting parts (e.g. colour coding or shape coding). If asymmetric parts are used, it may be advantageous to exaggerate the asymmetry to improve visual cues (Chhabra and Ahluwalia, 1990). Colour coding of parts

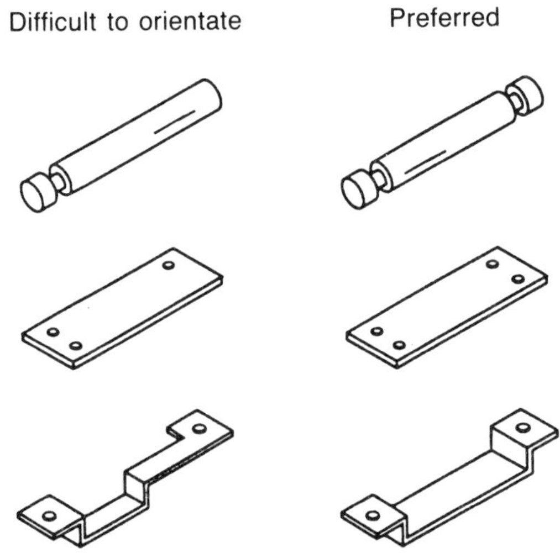

Figure 17.9 Use parts that are easy to orientate, such as symmetrical parts

may be used to form *families of parts*, i.e. parts which belong together in a subassembly. Colour coding will enhance stimulus-response compatibility in assembly. This results in reduced reaction time and better eye-hand coordination (Figure 17.10).

3. Consider feeding parts. The use of vibratory bowl feeders or other types of electromechanical feeder simplify the presentation and grasping of parts (Figure 17.11). Alternatively, magazines for parts or trays of parts can be used by the operator. These devices were conceived for use in automated assembly. However, they are equally practical for manual assembly.

17.2.5 Facilitating Assembly

1. Use self-locating parts. Parts with chamfers, notches and guides for self-location simplify assembly (Figure 17.12). The use of chamfers, for example, reduces the amount of manual precision required to insert the part. (The insertion time with and without chamfers can be modelled using Fitts' law (Fitts and Posner, 1973).
2. Reduce tolerances in part mating. Figure 17.13 illustrates how a slotted hole may be used to simplify positioning and relax accuracy requirements.

17.2.6 Consideration of Stability and Durability

Parts that are weak or easily bent are difficult to assemble (Figure 17.14). These parts often cause extra work in quality control, visual inspection, and replacement. Grossmith (1992) noted that many

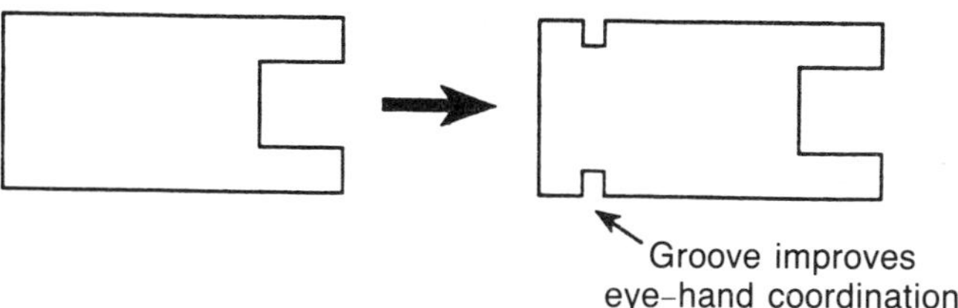

Figure 17.10 Exaggerated asymmetry may enhance stimulus-response compatibility

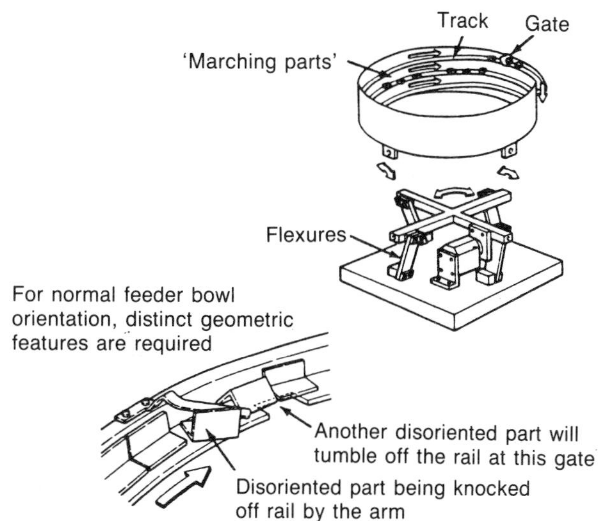

Figure 17.11 Use of a vibratory bowl feeder simplifies manual (and automatic) grasping of parts

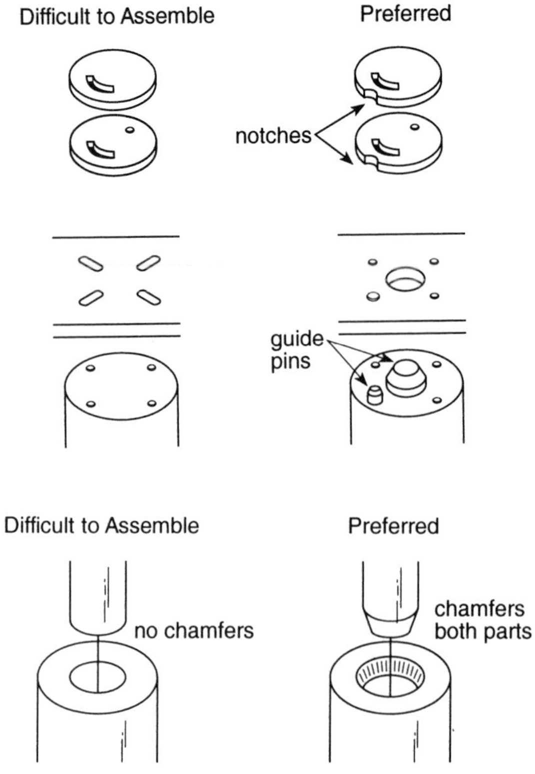

Figure 17.12 Facilitate assembly by using self-locating parts

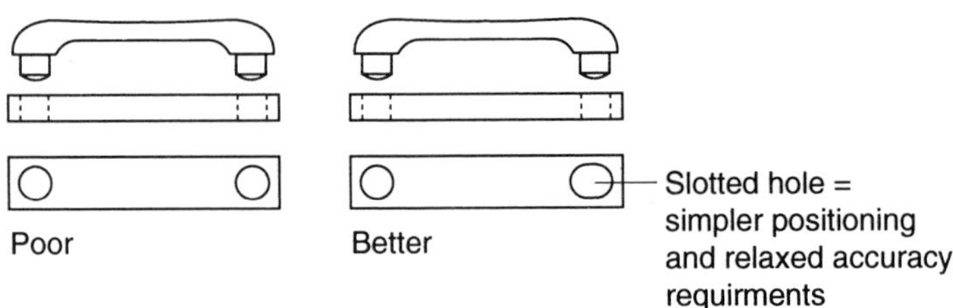

Figure 17.13 Reduce tolerances in part making

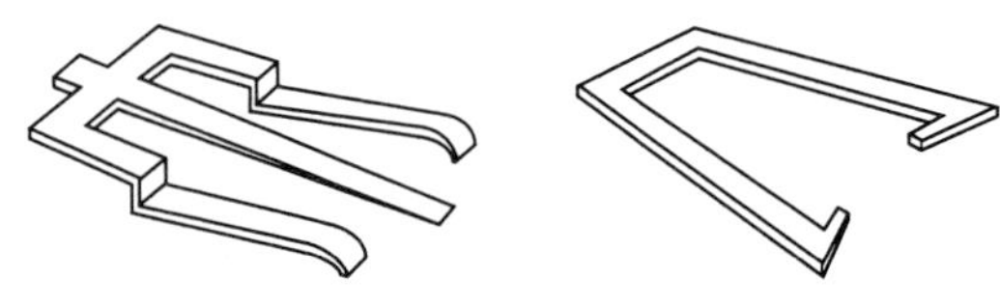

Figure 17.14 Avoid parts that are easily bent or parts that crack or chip

microscope inspection tasks can be avoided if product designers chose materials that are less likely to chip or crack.

17.3 Designing Automation Using Boothroyd's Principles

The design principles formulated by Boothroyd and Dewhurst (1983, 1987) have been extremely influential in industry. Several companies, including Hitachi, Black & Decker, General Electric, General Motors, IBM and Xerox, have used these principles to develop corporate guidelines (Gager, 1986; Holbrook and Sackett, 1988).

In Boothroyd's technique an existing product is disassembled. The necessity of each part is then analysed. First one must decide if a part is necessary for assembly or disassembly. If not, it may be possible to eliminate a part or integrate it with a mating part if:

1. There is no relative motion between the two parts.
2. The materials of which the two mating parts are composed do not have to be different.

For each part the assembly time is measured. Boothroyd then makes the assumption that an 'ideal' time for a part is 3 s. This is reasonable for a part that is easy to handle and insert. A measure of the manual assembly design efficiency (E_m) is then obtained using the equation:

$$E_m = \frac{3N_m}{T_m}$$

where N_m is the minimum number of parts, and T_m is the total assembly time. If $E_m < 1$ then the design is inefficient, and if $E_m > 1$ the design is efficient. An example of this methodology is given in Figure 17.15 and Table 17.1.

The value of E_m is, however, not always conclusive. Complex electromechanical products that require extensive wiring tend to have low design efficiencies, even when well designed. On the other hand, simple products with few parts can have a high design efficiency. In their handbook, Boothroyd and Dewhurst (1987) provide many examples of successful redesigns where productivity gains of 200–300% were obtained.

17.4 MTM Analysis of an Assembly Process

Boothroyd's technique is useful for redesigning existing products, but it cannot be used in the design of new products at the conceptual stages of design. Predetermined time-and-motion studies (PTMS) can be used for this purpose. As a basis for our analysis, we use motion time measurement (MTM) (e.g. Konz, 1990).

In MTM, an assembly is broken down into several constituent tasks including: reach, grasp (pick-up and select), move, position part, and insert. MTM specifies the amount of time it takes for a trained worker to do each of these elemental tasks. However, the asembly time depends very much on how the product is designed. Table 17.2 illustrates time savings for a 'best design case' as compared with less efficient designs. For example, reaching to a fixed location is the 'best case' and takes about 30% less time than reaching to a variable location or to small and jumbled parts. 'Grasping' of easily picked up parts is 75% faster than for parts that are not easily grasped.

Hence the design engineer should design parts that are easily reached and easily grasped. Luszack (1993) presented several methods which simplify the grasp of a part to be assembled

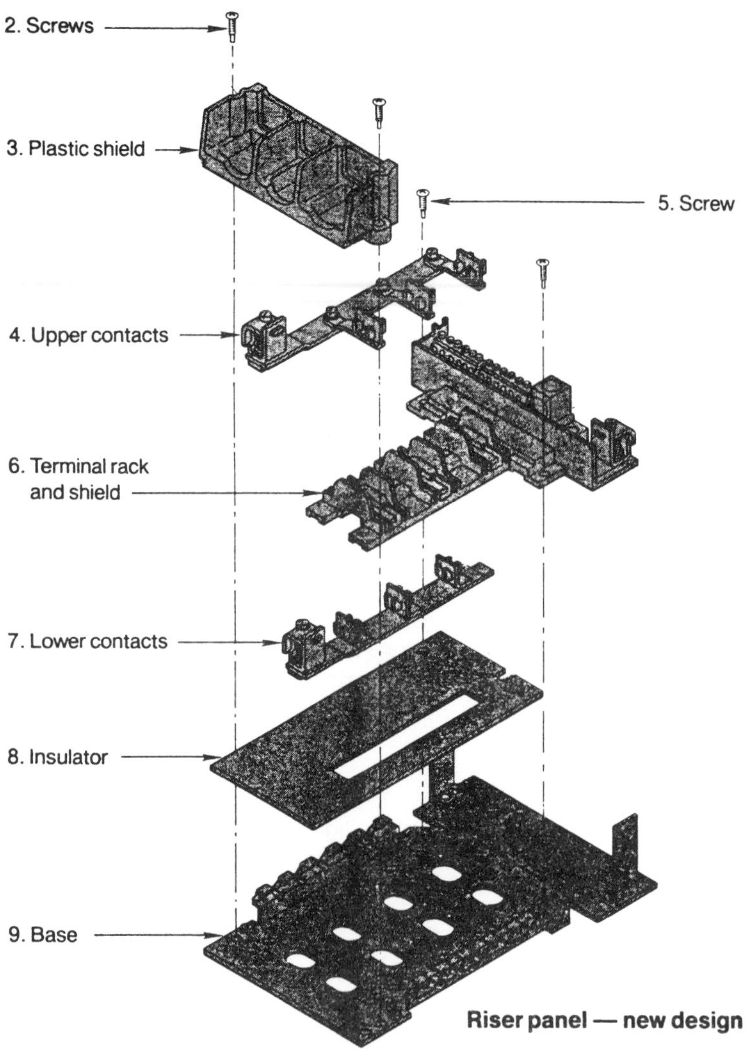

Figure 17.15 Example of product redesign for ease of assembly. The new design of the riser panel has only 10 parts, compared with 18 in the old design. Further reduction would be possible if screws 2 and 5 could be replaced by integral fasteners. The calculated assembly times are given in Table 17.1

(Figure 17.16). This is a complementary approach. Here, the parts are not redesigned to be easier to grasp, rather the process (of grasping) is redesigned. The process is redesigned for the purpose of improving the interactivity of products and processes so that (in this case) manual assembly will be easier. Process design is typically more abstract than product design, and hence more difficult to implement.

The parts should be presented at a fixed location. This can be accomplished by using part feeders (Figure 17.11). Much research has been performed to develop part feeders for robots (Boothroyd, 1982). These can also be used for manual assembly. A cost-benefit calculation can easily determine whether parts feeders for a manual assembly are cost-efficient. Simply calculate the time savings for assembly and compare to the cost for parts feeders.

Following the 'pick-up' the part has to be transported and positioned for the final insertion step. Table 17.2 illustrates that moving a part against a stop (case A) requires about 15% less time than when a

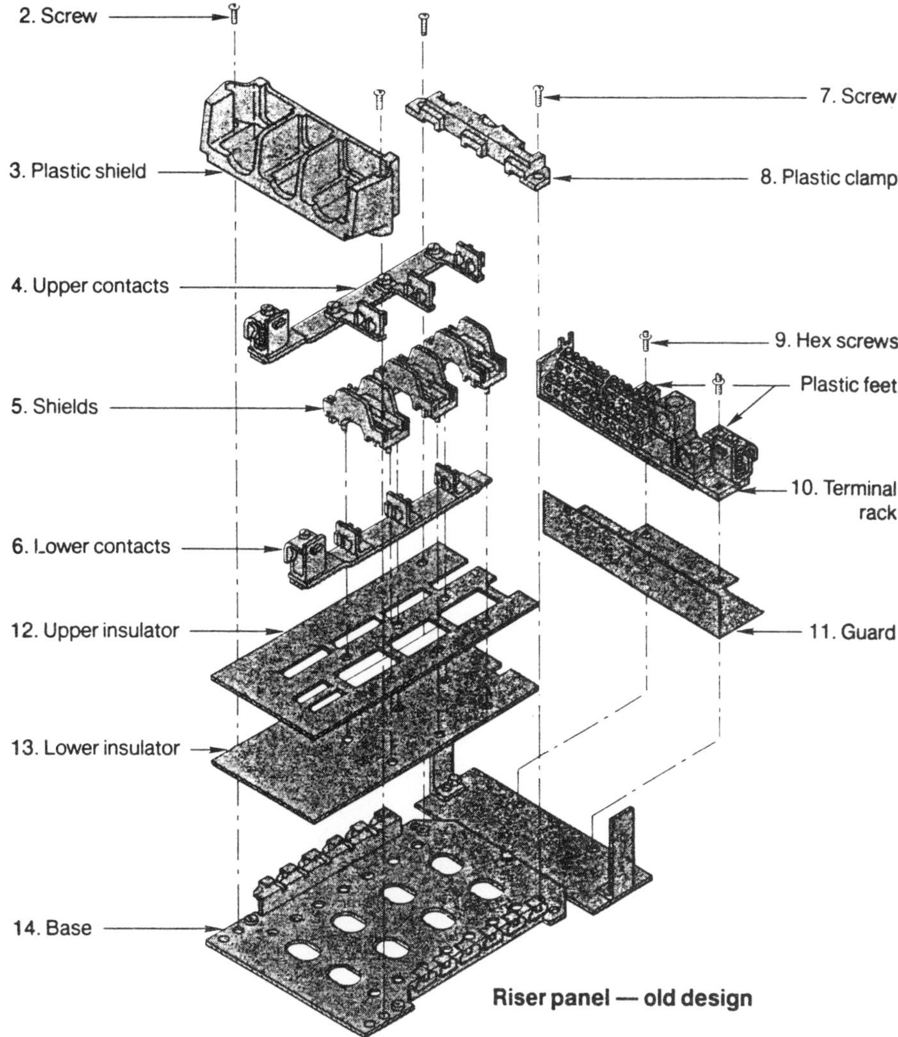

Figure 17.15 cont.

part is moved to a location without a stop (case B). In the latter case the absence of tactile feedback requires greater manual control. Ironically, most products are assembled as in case B. One objective of good design must therefore be to incorporate stops which provide tactile feedback (Furtado, 1990).

In MTM, parts insertion or mating is described using a position element composed of three complex motions; align, orientate and engage. 'Align' is the time required to line up the insertion axes of the two parts, like a pen into a cap. 'Orientate' describes the basic motions required to geometrically match the cross-sections of the two parts, like a key into a lock. 'Engage' consists of motions required to insert a part. Alignment is effected by asymmetry of the part. Table 17.2 illustrates that the use of symmetrical parts gives a time saving of 20%.

17.5 Human Factors Principles in Design for Assembly

In MTM and other PTMS methods we predict the time for manual assembly. These methods do not consider the time required for information processing. Yet there are many design features that can affect the information processing time. Table 17.3 lists several human factors principles that are applicable to design for human assembly

Table 17.1 Worksheet for design for manual assembly; the old design had 18 parts and required 136.3 s for assembly; the improved design has only 10 parts with an estimated assembly time of 51.7 s

1 Part ID No.	2 Number of times the operation is carried out consecutively	3 Two-digit manual handling code	4 Manual handling time per part	5 Two-digit manual insertion code	6 Manual insertion time per part	7 Operation time (s): (2)x[(4) + (6)]	8 Operation cost ($): 0.4x(7)	9 Figures for estimation of theoretical minimum parts	Name of assembly
Old design									
14	1	30	1.95	00	1.5	3.45	1.38	1	Base
13	1	33	2.51	00	1.5	4.01	1.60	1	Lower insulator (<2 mm)
12	1	33	2.51	06	5.5	8.01	3.20	0	Upper insulator (<2 mm)
11	1	30	1.95	08	6.5	8.45	3.38	0	Guard
10	1	30	1.95	08	6.5	8.45	3.38	1	Terminal rack
9	2	10	1.50	49	10.5	24.00	9.60	0	Hex, screws (8x16 mm)
8	1	30	1.95	00	1.5	3.45	1.38	0	Plastic clamp
7	2	11	1.80	49	10.5	24.60	9.84	0	Screw (9x14 mm)
6	1	30	1.95	08	6.5	8.45	3.38	1	Lower contacts
5	3	20	1.80	00	1.5	9.90	3.96	0	Shields
4	1	30	1.95	02	2.5	4.45	1.78	1	Upper contacts
3	1	30	1.95	02	2.5	4.45	1.78	1	Plastic shield
2	2	11	1.80	49	10.5	24.60	9.84	0	Screw (9x14 mm)
						$T_m = 136.3$	$C_m = 54.5$	$N_m = 6$	Design efficiency $= \frac{3N_m}{T_m} = 0.13$
New design									
9	1	30	1.95	00	1.5	3.45	1.38	1	Base
8	1	30	1.95	00	1.5	3.45	1.38	1	Insulator (>2 mm)
7	1	30	1.95	00	1.5	3.45	1.38	1	Lower contacts
6	1	30	1.95	00	1.5	3.45	1.38	1	Terminal rack and shield
5	2	10	1.50	38	6.0	15.00	6.00	0	Screw (9x20 mm)
4	1	30	1.95	02	2.5	4.45	1.78	1	Upper contacts
3	1	30	1.95	00	1.5	3.45	1.38	1	Plastic shield
2	2	10	1.50	38	6.0	15.00	6.00	0	Screw (9x20 mm)
						$T_m = 51.70$	$C_m = 20.68$	$N_m = 6$	Design efficiency $= \frac{3N_m}{T_m} = 0.35$

(DHA), including design features that reduce human information processing time.

Much of this chapter has addressed *ease of manipulation*, which will enhance productivity in manual assembly.

An example of *tactile feedback* is the use of physical stop barriers. When a part is moved against a stop there is a sensation in the fingers - tactile feedback which indicates that the task has been completed. *Auditory feedback* is helpful not only with parts but also for hand tools and controls and for hand tools operating on parts. In this case a sound is produced that indicates task completion. For example, the clicking sound of a switch, or the ricketing noise of a hydraulic screwdriver, indicating that the task was completed.

Visibility and *visual feedback* play an important role in assembly. Everything that is used in the manufacturing task should be fully visible. Hidden or invisible parts, cannot be pointed at. They become difficult to think of and are more abstract. When a task has been completed, there should be visual feedback - in other words something should look different. Sometimes in automobile assembly a piece of tape is put on top of a part to indicate it is finished.

Spatial compatibility has to do with the spatial layout of a workstation and has been addressed previously (see Figure 11.4). Part bins can be located in sequential order so that the operator can pick parts from

Table 17.2 Examples of time savings obtained with the 'best case' as compared with less efficient alternatives

	Best case	Comparison	Approximate time saving for best case (%)
Reach	To fixed location (case A)	To variable location (case B)	30
		To small or jumbled objects (case C)	40
Grasp			
Pick-up	Easily grasped (case 1A)	Object on flat surface (case 1B)	75
		Small object, 1/2 in. (case 1C2)	400
Select (for jumbled objects only)	Large jumbled objects (case 4A)	Object smaller than 1x1x1 in. (cases 4B, 3C)	50
Move	Against a stop or to other hand (case A)	To exact location without a stop or physical barrier (case C)	15
Position part			
Symmetrical part	Symmetrical, e.g. round peg in round hole (code S)	Semi-symmetrical 45° turn typical (code SS)	20
		Non-symmetrical, 75° turn typical (code NS)	30
Depth of insertion	No depth	4 in. insertion	100
Pressure to fit	Gravity, no pressure (code 1)	Light pressure (code 2)	210
		Heavy pressure (code 3)	500
Disengage (two parts)			
Class 1 fit	Loose	Tight	500
Ease of handling	Easy	Difficult	40

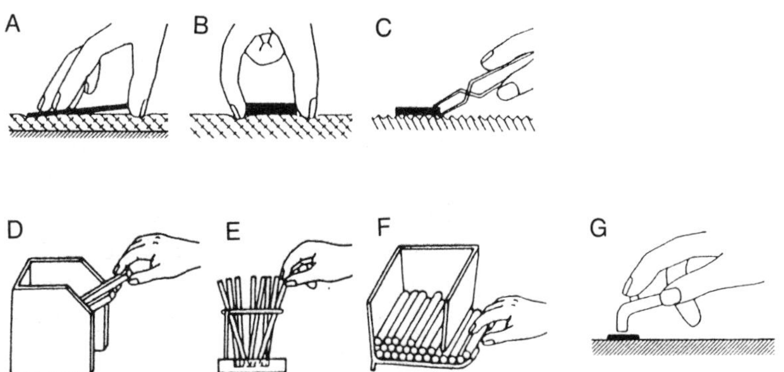

Figure 17.16 Gripping aids for an assembly workstation. (A and B) Gripping against a soft surface. (C) Tweezers or tongs used against a rippled table surface. (D) Container with inclined opening. (E) A ringholder with smaller bottom diameter. (F) Self-feeding container. (G) Use of vacuum gripper. From Luczak (1993)

Table 17.3 Human factors principles in DHA

Design for ease of manipulation
Design for tactile and auditory feedback
Design for visibility and visual feedback
Design for spatial compatibility
Design to enhance the formation of a mental model
Design for transfer of training
Design for job satisfaction

left to right in the same order as used in the assembly. Part bins can also be arranged so that their location mimics the product design. This could, for example, be used with components that are inserted in an electronic board. The best arrangement depends on the product design and the number of parts used. Obviously product design should consider spatial compatibility. One should also consider the locations of hand tools and controls. Typically items that belong together in task execution should be physically close.

Workers develop *mental models* of the task they are performing, i.e. they think of an assembly in a certain way. The concept of mental models has been used extensively in human–computer interaction. Software programmers have a different mental model than do users of the same software. Therefore, programmers fail to consider the needs of the user. Similarly, in manufacturing the product designer may fail to consider mental models other than his or her own. There are, indeed, many different tasks that impose different mental models (Baggett and Ehrenfeucht, 1991). A person assembling a product would have a different mental model than a person responsible for the quality control of the same product. They look for different things and they do different things, and the priorities are different. This observation is contrary to the notion that assembly operators should exercise their own quality control; it may be difficult to change a person's mindset (Shalin *et al*., 1994).

Transfer of training applies when a new product has only small modifications compared with the old product. A worker can then apply his skills to the new product. However, differences in product design and workstation layout may create confusion, and assembly times can increase drastically. Product designers have a responsibility here to make the assembly of new products similar to the assembly of previous products.

Design for job satisfaction is probably the most difficult aspect in planning for manufacturing. One problem is that people have different needs and are satisfied by different factors. We may understand better what factors lead to job dissatisfaction, and it could be easier to 'design to avoid job dissatisfaction'. More research is needed here.

Designers of manufacturing processes, facilities, and products must evaluate their design from the point of view of job satisfaction. There are several criteria (Locke, 1983). The design of a job should allow operators to:

- Collaborate.
- Talk to others.
- Receive performance feedback.
- Have control over their own work pace.

- Use their own judgement and decision-making.
- Be exposed to opportunities to learn new concepts and develop new skills.

These factors are affected by engineering design and should be addressed in the design process.

17.5.1 Example: Design for Job Satisfaction

Discuss the effects of product design and facilities design on job satisfaction. In particular, address the factors listed above. Provide examples of scenarios where these factors were not considered and where they were considered. For these scenarios, discuss what you think the effects will be on job satisfaction and job dissatisfaction.

Is it easier to predict when an individual will be satisfied or when an individual will be dissatisfied? Is there a difference in the types of issue that lead to satisfaction versus those that lead to dissatisfaction?

Chapter 18

Design for Maintainability

With increased complexity in manufacturing and use of computers and automated devices, maintenance is becoming more difficult. To maintain an automated piece of equipment or a robot, an operator needs knowledge of electronics, hydraulics, pneumatics and programming. In a manufacturing plant there is also an increased use of specialized machines or one-of-a-kind machines. In this complex scenario, it is important that production equipment is designed from the very start with maintainability in mind. To avoid expensive downtime, production equipment must be easy to maintain and quick to service. The design of equipment then becomes extremely important, since machines that are designed with maintainability in mind can effectively reduce the amount of downtime.

We need only look at the military scenario to understand that increased complexity of machines has a severe outcome on machine availability. Bond (1986) mentioned that, at any given moment, only about one-half of the combat aircraft on a US Navy carrier are able to fly off the ship with all the systems in 'up' condition. From military maintenance reports it appears that an average of 23% of all 'faulty' components sent back from the field for repair were actually in working condition. These components accounted for 20% of maintenance man-hours.

Below we discuss four aspects of maintainability: fault identification, testability and troubleshooting, accessibility, and ease of manipulation (Figure 18.1).

18.1 Ease of Fault Identification

Equipment should be designed so that it is easy to identify faults. From our perspective there are two interesting aspects:

- Increased use of diagnostic aids and software and automatic test equipment (ATE).
- Reduced complexity of machine design to simplify human fault identification.

Increasingly, ATE and other diagnostic tools are used in maintenance. Often, however, ATE is not helpful. Coppola (1984) reported that even in a well-seasoned system such as the AWACS surveyance aircraft the built-in test (BIT) capabilities were not very helpful. Out of 12 000 trouble indications, it turned out that 85% were

Ease of fault identification	Testability troubleshooting	Manual and visual accessibility	Ease of manipulation

Figure 18.1 The four steps of design for maintainability partly overlap

false alarms. Only 8%, or about 1000, of the incidents could be counted as real problems. About 6%, or 760 incidents, were confirmed problems, and BIT detected nearly all of them (98%). However, ATE could identify the reasons for only about half of the 760 failures, and the remaining 350 failures had to be investigated by maintenance technicians. This situation is typical of ATE: many false alarms, and much troubleshooting done by people. In fact half of the maintenance effort is typically taken by the '5% troubles' (Maxion, 1984). These are the difficult faults that elude automatic test routines, and experienced technicians are necessary to correct the faults.

ATE has become popular in civilian applications too. Xerox 1590 is an advanced copy machine with many built-in functions. This machine has built-in sensors which are monitored by a modem that is connected to a telephone line. Data are then transmitted at regular intervals from the copy machine to a central database, where the data are analysed remotely. Differences in data are used to determine malfunction of the equipment. The Kodak 300 copier has a similar system with an advanced communications facility and automatic detection of errors. However, there may still be problems with false alarms; the copy machines may not be any better than the ATE used with AWACS.

A complementary, and probably better approach is to design the new equipment with maintainability in mind. There could be ways of reducing the complexity of the equipment. In general, we want to design the equipment so that it is possible for the maintenance technician to 'chunk' the different components. Due to the limitations in short-term memory, it is difficult to think of individual components. It is much easier to think of modularized or functional blocks.

The maintenance technician will use systems charts for fault identification. Surprisingly, in the US Navy the basic system charts are often developed by human factors professionals (Bond, 1986). A good chart is typically more pictorial, hierarchical, and 'chunkable' than the usual information given in schematic diagrams.

Once the critical functionalities and components have been identified, there is still the problem of finding one's way inside the equipment. Here, we can offer concrete advice when it comes to labelling and colour coding of the various functional elements. Typically, these measures help in the identification of components as well as in determining the functional relationships between components. Tables 18.1 and 18.2 suggest some ways of using labelling and colour coding.

Table 18.1 Labels can be used in equipment to simplify maintenance (adapted from VanCott and Kinkade, 1972)

- Label access ports with information about components that can be reached through them
- Use labels to identify test points and present critical information. Use short and clear messages
- If any fasteners are not familiar or not common, label them to indicate how they should be used
- Use labels to identify potential hazards. Make the labels apparent to the casual operator
- Place the labels where they will not be destroyed by dirt, or wear

Table 18.2 The suggested use of colour coding in military environments has largely been adopted by industry

Colour	Indicates	Application
Red	Emergency or danger condition	Mainline power on
	Warns of energized or unsafe condition	Main breaker on
	Malfunction requiring maintenance	
	Stop	
	(Flashing red used when application requires a more compelling alert than a steady light)	
Amber or yellow	Motors running, machine in cycle	Test
	Alerts to condition requiring response, but not necessarily maintenance	Attention. Stand-by Intervention required
	(Flashing yellow used when application requires a more compelling alert than a steady light)	
Green	Safe condition	Start
	Go. Ready. Proceed	Cycle complete
	Go or start condition	Maintenance mode
White or clear	Major power not on	
	Normal conditions	
	Equipment operating conditions	AC on, AC off, power off, and auto-select
	Normal indications which do not have right, wrong, or alert significance	

Colour coding can be used in functional diagrams of equipment. Diagrams can be attached to the inside of a cover door; The choice of different colours will affect the perception of the state of a function. Thus, red is typically associated with 'stop', 'danger' and 'hot'. yellow is typically associated with 'caution' and 'near', whereas green is associated with 'go' (equipment operating in a normal fashion) and 'on'. Table 18.2 presents coding as used in military guidelines (Van Cott and Kinkade, 1972).

18.2 Design for Testability and Troubleshooting

Most real troubleshooting activity seems to be of 'opportunistic' nature, i.e. the technician will typically not plan ahead of time or generate a list of possible faults. Instead, he or she will act according to what it seems to be convenient to test. Testing points that are easily visible, and where the logic is clear, are likely to be tested first (Kieras, 1984). One of the best things an equipment designer can do for the maintenance person is to separate logically the different units of a piece of equipment. If this is done, a relatively simple check sequence can be used to decide what part of the equipment is faulty.

The goal of the design should be to provide a situation where the maintenance person has about the right amount of intellectual challenge to keep him or her interested. This can indeed be realized, as mainframe computer manufacturers have shown. Failures in large mainframes are usually repaired smoothly and rapidly, and without the need for creative problem solving in every case (Bond, 1986), yet there is still room for individual problem solving and a sense of competency and job satisfaction.

Bond (1986) has pointed out that the education of maintenance technicians is misconceived. For many decades electronics schools have started their students out by teaching them physics and basic electronics theory, on the grounds that teaching and learning these subjects is necessary for effective maintenance work. Actually, training in fundamentals or principles is of little use in maintenance, and most of the material is quickly forgotten (Morris and Rouse, 1984).

18.3 Design for Accessibility

In design for accessibility we are interested in two aspects: visual accessibility and reach accessibility. To enhance the visibility of components, it may be desirable to install lamp fixtures inside machines. It is also important to make the openings large and prominent and to locate the service components in a prominent location where they are easy to reach.

Maintenance is often performed using access ports to the equipment. Several issues should be considered when designing such ports (Table 18.3). These issues clearly matter. In one of the few experiments ever performed on maintainability, Kama (1963) demonstrated the effect of accessibility on work time (Figure 18.2). The US Department of Defense has taken much interest in accessibility, and many guidelines have been devised. Figure 18.3 shows the clearances for the hands necessary for equipment maintenance.

18.4 Design for Ease of Manipulation

To simplify maintenance one must consider the design of the components that are used in the equipment. Several guidelines for ease of manipulation of connectors and couplings are given in Table 18.4.

18.5 Summary

Design for maintainability is extremely important for concurrent engineering. Nonetheless, there has been very little research done that is directly applicable to manufacturing. Some companies are investigating related issues such as design for serviceability, which has become important in, for example, the design of copier machines. In these cases there is no fault finding, and diagnosis and design for serviceability present less of a challenge than design for

Table 18.3 Design of access ports for maintenance (adapted from Van Cott and Kinkade, 1972)

- Consider the requirements of the maintenance task in terms of tool use, exertion of force, and depth of reach. Use this information to determine the dimensions of access ports
- Provide openings to components that need maintenance. Openings must be large enough to permit access by both hands. Openings must also offer visibility of components
- Locate access ports so that they do not expose maintenance operators to hot surfaces, electrical currents or sharp edges
- Locate access ports so that the operator can monitor necessary display(s) while making adjustments

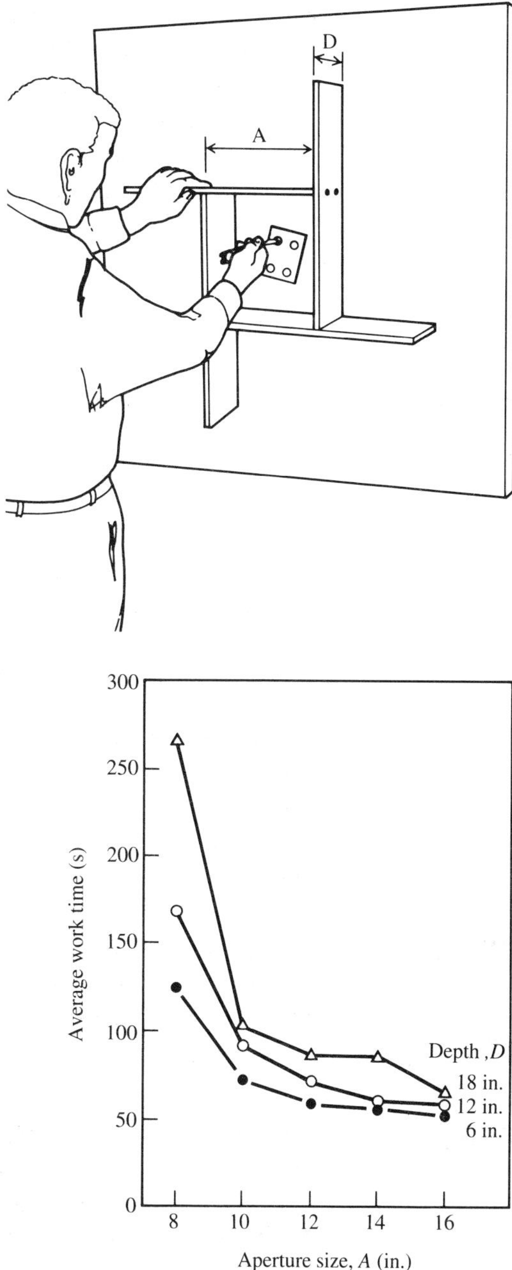

Figure 18.2 The width of the opening (A) and the depth of the opening (D) affect the average work time required for this maintenance task. Note that for aperture sizes less than 10 in. there is a dramatic increase in work time, and even more so for a depth of 18 in. than 6 in.

Table 18.4 Design of connectors and couplings to ease manipulation in maintenance (adapted from VanCott and Kinkade, 1972)

- Provide access ports that are easy to remove - if possible hinged
- Design fasteners for covers so that they are easily visible and accessible
- Fasteners on access covers should be easy to operate with gloved hand, e.g. tongue-and-slot design
- Minimize the number of turns necessary to remove components
- Use hexagonal bolt screws that can be removed using either a screwdriver or a wrench
- Make replaceable seals visible - to ensure that they are replaced

	Task
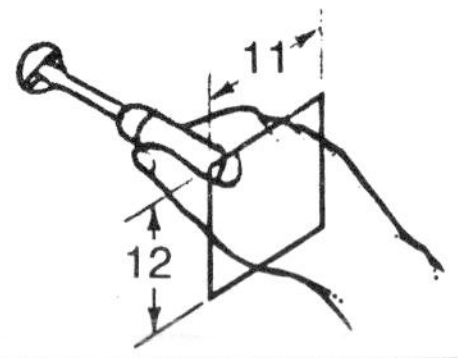 11, 12	Using common screwdriver, with freedom to turn hand through 180
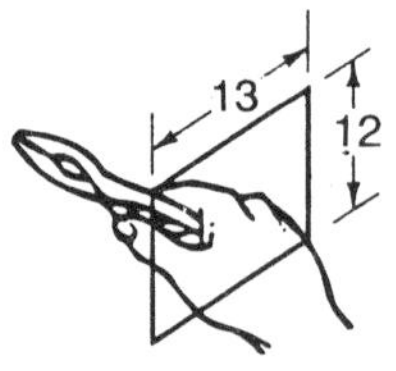 13, 12	Using pliers and similar tools
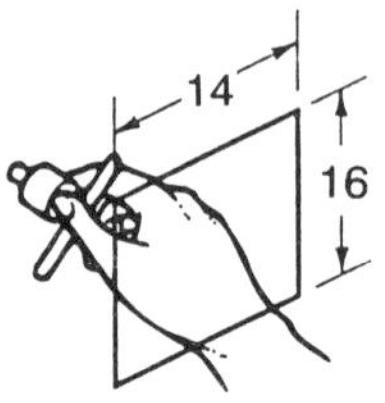 14, 16	Using "T" handle wrench, with freedom to turn hand through 180
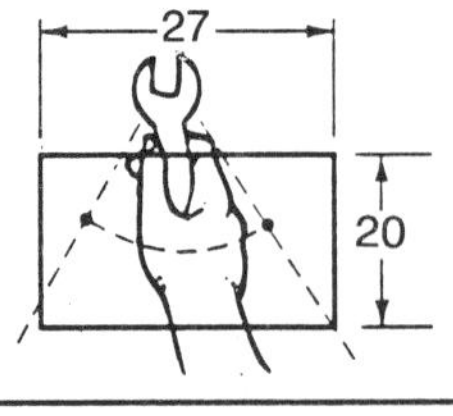 27, 20	Using open-end wrench,with freedom to turn wrench through 60
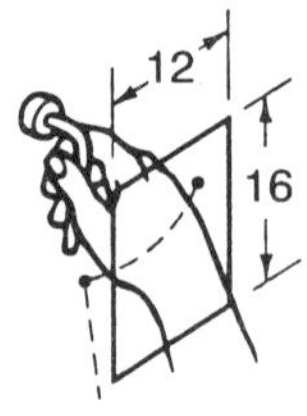 12, 16	Using Allen-type wrench with freedom to turn wrench through 60
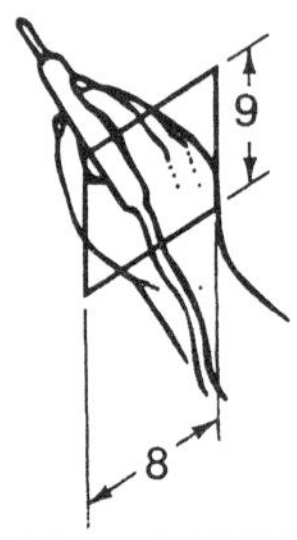 9, 8	Using test probe. etc.

Figure 18.3 Clearances (in cm) required for the hands. Additional information is available in Van Cott and Kinkade (1972) and US Military Standard 1472D (US Department of Defense, 1988)

maintainability. In design for serviceability it is important to present a coherent picture of the equipment. All common service tasks should be easy to perform and they should obvious to a user.

Finally, we should emphasize that operator comfort and convenience are paramount in design for maintainability. When maintaining a piece of equipment, it should be possible to adopt a natural work posture, such as sitting. Thus, the product designer should consider positioning common maintenance items at a comfortable work height.

Chapter 19

Machine Safety and Robot Safety

In this chapter we explain how different safety devices can be used to protect workers around machines. In the last couple of years, robot safety has been addressed to a greater extent than other types of machine safety, and so we focus on robotic safety here. However, many of the safety devices suggested below were developed for applications other than robotics, and are well suited to other types of machine.

There are many reasons to be concerned about robotic safety. Early statistics show that robots are more unsafe than punch presses, which previously were the most unsafe machines in industrial environments. An accident analysis showed that there are differences in the causation patterns between fatalities and injuries (Helander, 1990). Most fatalities occurred with heavy robots, such as manual handling robots, while the machine was operating. Typically, the worker had entered the robot area to do maintenance work. The robot then hit the operator from above or from behind, where the operator could not see it approaching, and the body or the skull of the operator was crushed.

Injuries inflicted by robots are different. In the first place they can be inflicted by much smaller robots, such as those used for assembly and welding. In most cases of injury the operator's hand was struck or pinned by the robot.

Safety authorities in many countries have published guidelines for robotics safety. The National Institute of Occupational Safety and Health (NIOSH) in the USA has published a safe maintenance guideline for robotic workstations (Etherton, 1988), and similar publications have been issued in Japan, the UK, Sweden and Germany.

19.1 Safety Devices

There are many different types of safety device that can be used to shut down a robot, in case the operator gets too close. Table 19.1 gives an overview of the most commonly used devices. The devices are divided into two categories: work area intrusion, and inside robot movement zone. Note that the last four sensors listed in Table 19.1 are still in the experimental stage, and currently they may not be practical to use.

19.1.1 Physical Barriers

Physical barriers such as fences, guard rails, chains, and curtains are used to prevent access to the working area. Fences are also used to protect people from flying objects that may be accidentally thrown by the robot. Depending upon the installation, fences can be designed to permit a flow of parts in and out of the working envelope. Access doors must be interlocked. In Figure 19.1, the workstation is divided into two zones: the robot movement zone, and the approach zone. In the robot movement zone, operators are exposed to the physical

Table 19.1 Current safety devices (note that the last four devices are at the developmental stage)

Safeguarding devices	Description	Typical application and restrictions
Work area intrusion		
Fencing	Guards with interlocked gates	Heavy material handling
Guard rails	Awareness barriers with interlocked gates	Light material handling
Chains and posts	Passive guard	Small assembly; light-duty applications
Curtain	Flexible screen	Protection from welding and heat
Photoelectric beams	Photocell/optical	Often used in combination with other devices
Pressure-sensitive mats and surfaces	On floor or other surface to sense walking or touching	Often used in combination with other devices
Floor markings	Painted floor warnings	Indicates robot work envelope
Inside robot movement zone		
Pressure-sensitive mats/surfaces/skin	Attached to robot or critical surrounding areas	Stops robot upon contact
Infrared sensor	Sensitive to infrared energy from person	Sensitive to ambient temperature as well
Camera image processing	Shape, size and contrast	Slow, due to computational response lag
Ultrasound sensor (sonar)	Sensitive to motion and distance	Directional (10–30°), temperature sensitive
Capacitance	Capacitive field changed by presence and distance	Potential problems with calibration and drift

power of the robot. In the approach zone there is danger from thrown objects, radiation and mechanical hazards. In this case, the operator can use a rotating worktable which, like a lazy Susan, is rotated into the work envelope.

Chains are passive guards that work as awareness barriers. They can be easily overcome by intruders. However, they may be appropriate for smaller robots used for light duty assembly, where their main function is to remind the operator.

The main purpose of a welding curtain is to protect operators from radiation and ultraviolet light from welding. They also have a secondary purpose, since they are in effect physical barriers.

19.1.2 Photoelectric Beams

Photo cells can be used for outlining the work barrier. Alternative devices such as light curtains and magnetic curtains are also sometimes used to detect personnel intrusion into the working envelope.

19.1.3 Pressure-sensitive Mats

These are used to detect a person walking towards the work envelope. Their use is impractical if parts have to be rolled in and out of the work envelope. Pressure-sensitive surfaces or skins may also be attached to a robot or the critical surrounding areas, so if an operator presses against or touches this area, the robot is stopped.

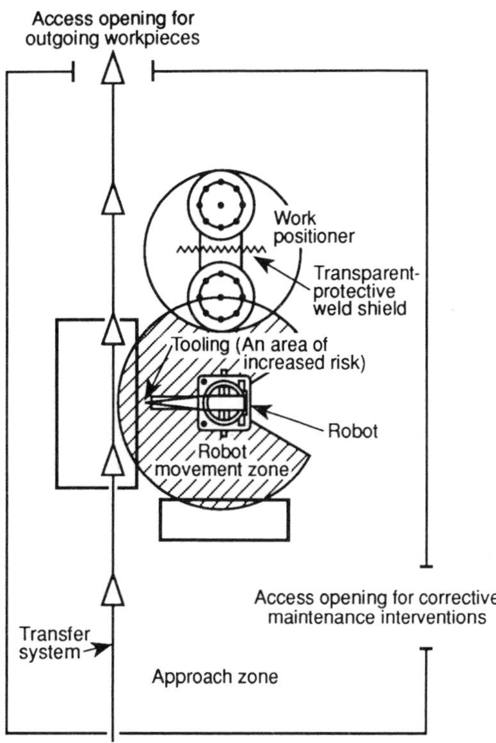

Figure 19.1 Safety design of a 'standard' robotic welding workstation according to the NIOSH (Etherton, 1988)

One version described by Baldur and Barron (1988) is a flexible strip which can be mounted on any curved surface. A contact force of 8 oz. will activate this device at any point along its length and will stop the robot.

19.1.4 Infrared Sensors

All objects in the environment emit infrared radiation. Infrared sensors have been developed that are tuned to the well-defined spectrum range of human infrared energy. Rahimi and Hancock (1988) have pointed out that many of these types of sensor can be used in hybrid workstations where operators and robots are cooperating.

19.1.5 Cameras and Image Processing

A camera image may be analysed using a pattern-recognition processor. By comparing consecutive images, it is possible to detect differences in the area, such as an entering operation.

19.1.6 Ultrasound (Sonar)

Ultrasound (or sonar) impulses are emitted, reflected against a human body, and then received again. There may be problems in using sonar technology for workstations. Jorgensen *et al.* (1986) have pointed out that, due to sensitivity to temperature changes, sonar sensors can make gross errors in the estimation of distance. False alarms may also be caused by multiple echoes or sonar signals emitted from other machines.

19.1.7 Capacitive Sensors

In this case a weak electromagnetic field is superimposed on the work area. An operator moving into the workplace changes the capacitance of the electromagnetic field. The capacitance is also affected by a number of disturbing factors such as changes in humidity, temperature, and many sources of electromagnetic noise.

19.2 Example: Case Study of Robot Safety at IBM Corporation/ Lexmark

The extensive safety precautions that surround robotic workplaces have actually been a deterrent to the development of automation, as this case study shows.

The IBM Corporation manufacturing plant in Lexington, Kentucky, specialized in printers and typewriters. During the 1980s an extensive system for automation was introduced. This was another 'engineer's dream' and most of the automation has since been discarded. At the beginning of the 1990s this plant was spun-off to form an independent company, Lexmark, which manufactures printers. One important reason for doing away with the robotic installation was the additional cost of robotic safety. Table 19.2 lists problems encountered at the IBM/Lexmark plant.

Although the automation for manufacturing was successful from a technological point of view, it was determined to be cost prohibitive. It was in fact cheaper and faster for some of the products to be assembled manually. Design for automation was considered impractical because of the short product life. For automation to be successful it must be possible to manufacture large volumes. Small-batch production is not effective. In the scenario with fairly small batches, manual intervention was needed for engineering changes and to solve quality problems. Under such circumstances, automation is not feasible.

19.3 Recommended Reading

Additional information about robotic safety is given in robotics safety standards issued by the Robotic Industries Association (1987, 1989). An excellent collection of papers on robotics safety has been edited by Rahimi and Karwowski (1993).

Table 19.2 Problems of automation and robotic safety at IBM Lexington/Lexmark

- Robot safeguards increase the cost of the automated manufacturing system
- Little space was left over after robot installation, causing robots to bump elbows. Entrapment became a concern
- Safety regulations did not consider maintenance of robots, where the needs were the greatest. They were designed for normal operation
- Robot workstations were difficult to change. Additional engineering updates usually involved major modifications of the robot system
- The line was down much of the time, necessitating an employee to 'hand assemble' the product. No provisions had been made for 'manual intervention stations' and thus numerous ergonomic problems, such as bad work postures, arose
- The constant movement of conveyors, robots and other automation produced noise levels that needed abatement
- The manufacturing lines 'backed-up' when the operator had to turn off a robot to enter the work envelope
- Operators began losing production and circumventing safety systems. Near-miss reports increased

References

Andersson, G.B.J. and Örtengren, R., 1974, Lumbar disc pressure and myoelectric back muscle activity during sitting, *Scandinavian Journal of Rehabilitation Medicine*, **3**, 115–121.

ANSI, 1986, *Octave-Band and Fractional Octave-Band Analog and Digital Filters, ANSI S1.11*, New York: American National Standards Institute.

ANSI, 1991, *Specifications for Personal Noise Dosimeters, ANSI S1.25*, New York: American National Standards Institute.

Armstrong, T.J. and Chaffin, D.B., 1979, Carpal tunnel syndrome and selected personal attributes. *Journal of Occupational Medicine*, **21**, 481–486.

Åstrand, I., 1969, Aerobic work capacity in men and women with special reference to age, *Acta Physiologica Scandinavica*, **49** (Suppl. 169).

Åstrand, P.-O. and Rodahl, K., 1986, *Textbook of Work Physiology*, New York: McGraw-Hill.

Ayoub, M.M., 1973, Workplace design and posture, *Human Factors*, **15**, 265–268.

Baggett, P. and Ehrenfeucht, A., 1991, Building physical and mental models in assembly tasks, *International Journal of Industrial Ergonomics*, **7**, 217–228.

Bailey, R.W., 1982, *Human Performance Engineering. A Guide for Systems Designers*, Englewood Cliffs, NJ: Prentice-Hall.

Baldur, R. and Baron, L., 1988, Sensors for safety, in Dorf, N. (Ed.) *International Encyclopedia of Robotics*, New York: Wiley.

Bennett, C., Chitlangia, A. and Pangrekar, A., 1977, Illumination levels and performance of practical visual tasks, *Proceedings of the Human Factors Society 21st Annual Meeting*, pp. 322–325, Santa Monica, CA: The Human Factors and Ergonomics Society.

Beranek, L. and Newman, R., 1950, Speech interference levels as criteria for rating background noise in offices, *Journal of the Acoustical Society of America*, **22**, 671.

Beranek, L., Balzier, W. and Figwer, J., 1971, Preferred noise criteria (PNC) curves and their application to rooms, *Journal of the Acoustical Society of America*, **50**, 1223–1228.

Bergqvist, U., 1986, Bildskärmsarbete och hälsa, in *Arbete och Hälsa*, Vol. 9, Stockholm: Arbetarskyddsverket.

Bergum, B.O. and Bergum, J.E., 1981, Population stereotypes: an attempt to measure and define, *Proceedings of the Human Factors Society 25th Annual Meeting*, pp. 662–665, Santa Monica, CA: The Human Factors and Ergonomics Society.

Bjerner, B. and Swensson, A., 1953, Shiftwork and rhythm, *Acta Medica Scandinavica*, **278**, 102–107.

Blackwell, H.R., 1964, Further validation studies of visual task evaluation, *Illuminating Engineering*, **59**, 627–641.

Blackwell, H.R., 1967, The evaluation of interior lighting on the basis of visual criteria, *Applied Optics*, **6**, 1443–1467.

Blackwell, O. and Blackwell, H., 1971, IERI report: visual performance data for 156 normal observers of various ages, *Journal of Illuminating Engineering Society*, **1**, 2–13.

Bond, N.A., Jr., 1986, Maintainability, in Salvendy, G. (Ed.), *Handbook of Human Factors*, pp. 1329–1355, New York: Wiley.

Boothroyd, G., 1982, *Design for Assembly Handbook*, Amherst, MA: Department of Mechanical Engineering, University of Massachusetts.

Boothroyd, G. and Dewhurst, P., 1983, *Design for Assembly*, Amherst, MA: Department of Mechanical Engineering, University of Massachusetts.

Boothroyd, G. and Dewhurst, P., 1987, *Product Design for Assembly*, Wakefield, RI: Boothroyd Dewhurst.

Boyce, P. R., 1981a, Lighting and visual performance, *International Review of Ergonomics*, **2**, London: Taylor & Francis.

Boyce, P.R., 1981b, *Human Factors in Lighting*, New York: Macmillan.

Braidwood, R., 1951, *Prehistoric Men*, Chicago, IL: Natural History Museum.

Branton, P., 1984, Backshapes of seated persons – how close can the interface be designed?, *Applied Ergonomics*, **15**, 105–107.

Broadbent, D., 1977, Language and ergonomics, *Applied Ergonomics*, **8**, 15–18.

Broadbent, D., 1978, The current state of noise research: reply to Poulton, *Psychological Bulletin*, **85**, 1052–1067.

Brogmus, G.E. and Marko, R., 1990, Cumulative trauma disorders of the upper extremities, *Proceedings of IEA Conference on Human Factors in Design for Manufacturability and Process Planning*, pp. 49–59, Santa Monica, CA: The Human Factors and Ergonomics Society.

Brown, A.C. and Brengelmann, G., 1965, Energy metabolism, in Ruch, T.C. and Patton, H.D. (Eds), *Physiology and Biophysics*, Philadelphia, PA: Saunders.

Brown, C.R. and Schaum, D.L., 1980, User-adjusted VDU Parameters, in Grandjean, E. and Vigliani, E. (Eds), *Ergonomic Aspect of Visual Display Terminals*, London: Taylor & Francis.

Browne, R.C., 1949, The day and night performance of the tele-printer switchboard operators, *Occupational Psychology*, **23**, 121–126.

Burri, G. and Helander, M.G., 1991a, A field study of productivity improvements in the manufacturing of circuit boards, *International Journal of industrial Ergonomics*, **7**, 207–216.

Burri, G. and Helander, M.G., 1991b, Implementation of human factors principles in the design of manufacturing process, in Pulat, B.M. and Alexander, D.C. (Eds), *Industrial Ergonomics Case Studies*, Norcross, GA: Industrial Engineering and Management Press.

Cakir, A., Hart, D.J. and Stewart, T.F.M., 1980, *Visual Display Terminals*, New York: Wiley.

Carlsson, L., 1979, *Ljus-och belysningskrav vid arbete med bildskärmar på tidningsföretag*, Stockholm: Tidningarnas Arbetsmiljö-Kommittee.

Carroll, J.M., 1993, Techniques for minimalist documentation and user interface design, in Jansen, C., Vander Poort, P., Steehouder, M. and Verhejen, R. (Eds), *Quality of Technical Documentation: Utrecht Studies in Language and Communication*, Vol. 2, Amsterdam: Rodopi.

Casali, J.G. and Park, M.Y., 1990, Attenuation performance of four hearing protectors under dynamic movement and different user fitting conditions, *Human Factors*, **32**, 9–25.

Chaffin, D.B., 1969, A computerized biomechanical model: development of and use in studying gross body actions, *Journal of Biomechanics*, **2**, 429–441.

Chaffin, D.B. and Andersson, G.B.J., 1991, *Occupational Biomechanics*, New York: Wiley.

Chapanis, A., 1971, The search for relevance in applied research, in Singleton, W.T., Fox, J.G. and Whitfield, D. (Eds), *Measurement of Man at Work*, pp. 1–14, London: Taylor & Francis.

Chapanis, A., 1974, National and cultural variables in ergonomics, *Ergonomics*, **17**, 153–176.

Chapanis, A., 1990, The International Ergonomics Association: its first 30 years, *Ergonomics*, **33**, 275–282.

Chapanis, A. and Kinkade, R.G., 1972, Design of controls, in Van Cott, H.P. and Kinkade, R.G. (Eds), *Human Engineering Guide to Equipment Design*, Washington, DC: US Government Printing Office.

Chapanis, A. and Lindenbaum, L., 1959, A reaction time study of four control-display linkages, *Human Factors*, **1**, 1–7.

Chhabra, S.L. and Ahluwalia, R.S., 1990, Rules and guidelines for ease of assembly, in *Proceedings of the International Ergonomics Association Conference on Human Factors in Design for Manufacturability and Process Planning*, pp. 93–99, Santa Monica, CA: Human Factors and Ergonomics Society.

Childe, G., 1944, *The Story of Tools*, London: Cobbet.

Cohen, H., 1979, *Conveyor Safety*, Washington, DC: US Department of HEW, NIOSH.

Collins, A.M., 1973, Decrements in tracking and visual performance during vibration, *Human Factors*, **15**, 379–393.

Collins, B. and Lerner, N., 1983, *An evaluation of exit symbol visibility, NBSIR Report 82-2685*, Washington, DC: National Bureau of Standards.

Coppola, A., 1984, Artificial intelligence applications to maintenance, in *Artificial Intelligence in Maintenance*, pp. 23–44, Brooks AFB, TX: Air Force Human Resources Laboratory.

Courtney, A.J., 1986, Chinese population stereotypes: color associations, *Human Factors*, **28**, 97–99.

Crossman, E.R.F.W., 1959, A theory of the acquisition of speed-skill, *Ergonomics*, **2**, 153–166.

Dedobbeleer, N. and Beland, F., 1989, Safety climate in construction sites, *Proceedings of International Conference on Strategies for Occupational Accident Prevention*, Stockholm: The National Institute for Occupational Health.

Deutsches Institut für Normung, 1981, *DIN Standard 66234, Parts 1 to 9*, Berlin: DIN.

Drillis, R.J., 1963, Folk norms and biomechanics, *Human Factors*, **5**, 427–441.

Drury, C.G., 1983, Task analysis methods in industry, *Applied Ergonomics*, **14**, 19–28.

Drury, C.G., 1991, Ergonomics practice in manufacturing, *Ergonomics*, **34**, 825–839.

Ducharme, R.E., 1973, Problem tools for women, *Industrial Engineering*, **Sept**., 46–50.

Eastman Kodak Co., 1983, *Ergonomic Design for People at Work*, Vol. 1, New York: Van Nostrand Reinhold.

Eastman Kodak Co., 1986, *Ergonomic Design for People at Work*, Vol. 2, New York: Van Nostrand Reinhold.

EC Council Directive L156, *Official Journal of the European Communities*, 1990, **21 June**, 9–13.

Eckstrand, G.A., 1964, *Current Status of the Technology of Training. Report AMRL-TDR-64-86*, Ohio: Aerospace Medical Laboratories, Wright–Patterson Air Force Base.

Elliott, T.K. and Joyce, R.D., 1971, An experimental evaluation of a method for simplifying electronic maintenance, *Human Factors*, **13**, 217–227.

Eriksson, R., 1976, Personal communication, International Labor Organization, Geneva, Switzerland.

Etherton, J., 1986, The use of safety devices and safety controls at industrial machine workstations, in Salvendy, G. (Ed.), *Handbook of Human Factors*, New York: Wiley.

Etherton, J.R., 1988, *Safe Maintenance Guidelines for Robotic Workstations*, Morgantown, WV: National Institute for Occupational Safety and Health.

Farrell, R.J. and Booth, J.M., 1984, *Design Handbook for Imagery Interpretation Equipment*, Seattle, WA: Boeing Aerospace Company.

Faulkner, T.W. and Murphy, T.J., 1973, Lighting for difficult visual tasks, *Human Factors*, **15**, 149–159.

Fellows, G.L. and Freivalds, A., 1991, Ergonomics evaluation of a foam rubber grip for tool handles, *Applied Ergonomics*, **22**, 225–230.

Fitts, P.M. and Posner, M., 1973, *Human Performance*, Englewood Cliffs, NJ: Prentice-Hall.

Fitts, P.M. and Seeger, C.M., 1953, S–R compatibility: spatial characteristics of stimulus and response codes, *Journal of Experimental Psychology*, **46**, 199–210.

Floderus, B., Stenlund, C. and Törnqvist, S., 1993, An update of recent epidemiological and experimental studies, in *Proceedings of Minisymposium on Electromagnetic Fields and Cancer*, Nice, France, Solna, Sweden: National Institute for Occupational Health.

Folkard, S., Monk, T.H. and Lobban, M.C., 1978, Short and long-term adjustment of circadian rhythms in 'permanent' night nurses, *Ergonomics*, **21**, 785–799.

Fothergill, L.C. and Griffin, M.J., 1977, The evaluation of discomfort produced by multiple frequency whole-body vibration, *Ergonomics*, **20**, 263–270.

Furtado, D., 1990, Principles of design for assembly, in *Proceedings of the International Ergonomics Association Conference on Human Factors in Design for Manufacturability and Process Planning*, pp. 147–52, Santa Monica, CA: Human Factors Society.

Gager, R., 1986, Design for productivity saves millions, *Appliance Manufacturer*, Jan. 46–51.

Garg, A. and Herrin, G.D., 1979, Stoop or squat? A biomedical and metabolic evaluation, *Transactions of American Institute of Industrial Engineers*, **11**, 293–302.

Garg, A., Chaffin, D.B. and Freivalds, A., 1982, Biomechanical stresses from manual load lifting: a static vs. dynamic evaluation, *Transactions of the American Institute of Industrial Engineers*, **14**, 272–281.

Gawron, V., 1982, Performance effects of noise intensity, psychological set, and task type and complexity, *Human Factors*, **24**, 225–243.

Genaidy, A.M., Duggai, J.S. and Mital, A., 1990, A comparison of robot and human performances for simple assembly tasks, *International Journal of Industrial Ergonomics*, **5**, 73–81.

Gilbert, T.F., 1974, On the relevance of laboratory investigation of learning to self-instructional programming, in Lamsdaine, A.A. and Glaser, R. (Eds), *Teaching Machines and Programmed Instructions*, Washington, DC: National Educational Association.

Goldstein, I.L., 1980, Training in work organizations, *Annual Review of Psychology*, **31**, 229–272.

Goldstein, I.L., 1986, The relationship of training goals and training systems, in Salvendy, G. (Ed.), *Handbook of Human Factors*, New York: Wiley.

Gould, J.D. and Grischkowsky, N., 1984, Doing the same work with hard copy and cathode-ray tube (CRT) computer terminals, *Human Factors*, **26**, 323–338.

Grandjean, E. (Ed.), 1984, *Ergonomics and Health in Modern Offices*, London: Taylor & Francis.

Grandjean, E., 1988, *Fitting the Task to the Man*, London: Taylor & Francis.

Grandjean, E., 1986, Design of VDT workstations, in Salvendy, G. (Ed.), *Handbook of Human Factors*, pp. 1359–1398, New York: Wiley.

Grandjean, E. and Vigliani, E. (Eds), 1980, *Ergonomic Aspects of Visual Display Terminals*, London: Taylor & Francis.

Greenburg, L. and Chaffin, D., 1977, *Workers and Their Tools*, Midland, MI: Pendall.

Greenstein, J.S. and Arnaut, L.Y., 1988, Input devices, in Helander, M. (Ed.), *Handbook of Human-Computer Interaction*, Amsterdam: North Holland.

Grether, W.F., 1971, Vibration and human performance, *Human Factors*, **13**, 203–216.

Grossmith, E.J., 1992, Product design considerations for the reduction of ergonomically related manufacturing costs, in Helander, M. and Nagamachi, M., *Design for Manufacturability. A Systems Approach to Concurrent Engineering and Ergonomics*, London: Taylor & Francis.

Gruber, G.J., 1976, *Relationships Between Wholebody Vibration and Morbidity Patterns Among Interstate Truck Drivers*, San Antonio, TX: Southwest Research Institute.

Hadler, N.M., 1986, Industrial rheumatology, *The Medical Journal of Australia*, **144**, 191–195.

Hadler, N., 1989, Personal communication, Denver, CO, USA.

Hagberg, M. and Sundelin, G., 1986, Discomfort and load on the upper trapezius muscle when operating a word processor, *Ergonomics*, **29**, 1637–1645.

Hale, A.R. and Glendon, A.I., 1987, *Individual Behavior in the Control of Danger*, Amsterdam: Elsevier.

Handbuch für Beleuctung, 1975, Essen: Verlag Girardet.

Hansson, J.E., Klussel, L., Svensson, G. and Winström, B.O., 1976, Working environment for truck drivers – an ergonomic and hygienic study, *Arbete och Hälsa*, Vol. 6, Stockholm: Arbetarskyddsstyrelsen.

Harris, W. and Mackie, R.R., 1972, *A Study of the Relationships Among Fatigue, Hours of Service and Safety of Operations of Truck and Bus Drivers (Technical Report 1727–2)*, Goleta, CA: Human Factors Research, Inc.

Hasselquist, R.J., 1981, Increasing manufacturing productivity using human factors principles, in *Proceedings of the 28th Annual Meeting of the Human Factors Society*, pp. 204–206, Santa Monica, CA: The Human Factors and Ergonomics Society.

Health and Safety Commission, 1991, *Handling Loads at Work – Proposals for Regulation and Guidance*, London: Health and Safety Executive.

Helander, E.A.S., 1992, Personal communication, World Health Organization, Geneva, Switzerland.

Helander, M.G., 1986, Design of visual displays, in Salvendy, G. (Ed.), *Handbook of Human Factors/Ergonomics*, New York: Wiley.

Helander, M.G., 1990, Ergonomics and safety considerations in the design of robotics workplaces: a review and some priorities for research, *International Journal of Industrial Ergonomics*, **6**, 127–149.

Helander, M.G. and Burri, G.J., Jr., 1994, Cost effectiveness of ergonomics and quality improvements in electronics manufacturing, *International Journal of Industrial Ergonomics*, in press.

Helander, M.G. and Domas, K., 1986, Task allocation between humans and robots in manufacturing, *Material Flow*, **3**, 175–85.

Helander, M.G. and Little, S., 1993, Preferred settings in chair adjustments, in *Proceedings of the Human Factors and Ergonomics Society 37th Annual Meeting*, pp. 448–454, Santa Monica, CA: The Human Factors and Ergonomics Society.

Helander, M.G. and Nagamachi, M., 1992, *Design for Manufacturability. A Systems Approach to Concurrent Engineering and Ergonomics*, London: Taylor & Francis.

Helander, M.G. and Palanivel, T., 1990, *Anthropometric Survey of Employees at IBM Corporation, San Jose (Confidential)*, Buffalo, NY: Ergonomics Research, Inc.

Helander, M.G. and Rupp, B., 1984, An overview of standards and guidelines for visual display terminals, *Applied Ergonomics*, **15**, 185–195.

Helander, M.G. and Schurick, J.M., 1982, Evaluation of symbols for construction machines, in *Proceedings of the 26th Annual Meeting of the Human Factors Society*, Santa Monica, CA: The Human Factors and Ergonomics Society.

Helander, M.G. and Waris, J.D., 1993, Effect of spatial compatibility in manual assembly of performance, in Marras, W.S., Karwowski, W., Smith, J.L. and Pacholski, L. (Eds), *The Ergonomics of Manual Work*, London: Taylor & Francis.

Helander, M.G., Billingsley, P.A. and Schurick, J.M., 1984, An evaluation of human factors research on visual display terminals in the workplace, *Human Factors Review*, **1**, 55–129.

Hickling, E.M., 1985, *An Investigation on Construction Sites as Factors Affecting the Acceptability and Wear of Safety Helmets*, Loughborough: University of Technology, Institute for Consumer Ergonomics.

Hill, S.G. and Kroemer, K.H.E., 1989, Preferred declination and the line of sight, *Human Factors*, **28**, 127–134.

Hocking, B., 1987, Epidemiological aspects of 'repetition strain injury' in Telecom, Australia, *The Medical Journal of Australia*, **147**, 218–222.

Holbrook, A.E.H. and Sackett, P.J., 1988, Design for assembly guidelines for product design, *Assembly Automation*, **8**, 202–211.

Holding, D.H., 1986, Concepts of training, in Salvendy, G. (Ed.), *Handbook of Human Factors*, New York: Wiley.

Hopkinson, R.G. and Collins, J.B., 1970, *The Ergonomics of Lighting*, London: Macdonald.

Horne, J.A. and Östberg, O., 1976, A self-assessment questionnaire to determine morningness–eveningness in human circadian rhythms, *International Journal of Chronobiology*, **4**, 97–110.

Hornick, R., 1973, Vibration, in *Bioastronautics Data Book*, 2nd edn, (*NASA SP-3006*), Washington, DC: National Aeronautics and Space Administration.

Human Factors Society, 1988, *ANSI/HFS 100. American National Standard for Human Factors Engineering of Visual Display Terminal Workplaces*, Santa Monica, CA: The Human Factors and Ergonomics Society.

Illuminating Engineering Society, 1982, *Office Lighting*, New York: ANSI/IES.

International Labor Office, 1987, *Checklist for Workplace Inspection for Improving Safety, Health and Working Conditions*, Geneva: ILO.

International Labour Organization, 1972, *Kinetic Methods of Manual Handling in Industry (Occupational Health Series No. 10)*, Geneva, ILO.

International Standards Organization, 1995, *ISO Series 9241*, Geneva: ISO.

International Standards Organization, 1976, *ISO 2631 Human Exposure to Wholebody Vibration*, Geneva: ISO.

International Standards Organization, 1984, *ISO 7730 Moderate Thermal Environments – Determination of the PMV and PPD Indices and Specification of the Conditions for Thermal Contrast*, Geneva: ISO.

International Standards Organization, 1989a, *ISO 7933 Hot Environments – Analytical Determination of Thermal Stress Using Calculation of Required Sweat Rate*, Geneva: ISO.

International Standards Organization, 1989b, *ISO 7243 Hot Environments – Estimation of the Heat Stress on Working Man, Based on the WBGT-Index (Wet Bulb Globe Temperature)*, Geneva: ISO.

Johansson, G. and Backlund, F., 1970, Drivers and road signs, *Ergonomics*, **13**, 749–759.

Jorgensen, C., Hamel, W. and Weisbin, C., 1986, Autonomous robot navigation, *BYTE*, **4**, 223–235.

Karlsson, K., 1989, *Bullerskador*, Stockholm: Arbetsmiljö-kommissionen.

Karwowski, W., 1991, Complexity, fuzziness, and ergonomic incompatibility issues in the control of dynamic work environments. *Ergonomics*, **34**, 671–686.

Kaufman, J. and Christensen, J. (Eds), 1984, *IES Lighting Handbook*, New York: Illuminating Engineering Society of North America.

Keegan, J.J., 1953, Alterations of the lumbar curve related to posture and seating, *The Journal of Bone and Joint Surgery*, **35A**, 589–603.

Keyserling, W.M. and Chaffin, D.B., 1986, Occupational ergonomics – methods to evaluate physical stress on the job, *American Review of Public Health*, **7**, 77–104.

Kieras, D.E., 1984, The psychology of technical devices and technical discourse, in *Artificial Intelligence in Maintenance*, pp. 227–254, Brooks AFB, TX: Air Force Human Resources Laboratory.

Kinkade, R.G. and Wheaton, G.R., 1972, Training device design, in Van Cott, H.P. and Kinkade, R.G. (Eds), *Human Engineering Guide to Equipment Design*, Washington, DC: US Government Printing Office.

Knauth, P., Rohmert, W. and Rutenfranz, J., 1979, Systematic selection of shift plans for continuous production with the aid of work-physiological criteria, *Applied Ergonomics*, **10**, 9–15.

Knauth, P., Kiesswetter, E., Ottman, W., Karvonen, M.J. and Rutenfranz, J., 1983, Time-budget studies of policemen in weekly or swiftly rotating shift systems, *Applied Ergonomics*, **14**, 247–252.

Kokoschka, S. and Haubner, P., 1985, Luminance ratios at visual display workstations and visual performance, *Lighting Research and Technology*, **17**, 138–145.

Konz, S., 1990, *Work Design: Industrial Ergonomics*, Worthington, OH: Publishing Horizons.

Konz, S., 1992a, Macro-ergonomic guidelines for production planning, in Helander, M. and Nagamachi, M. (Eds), *Design for Manufacturability*, pp. 281–300, London: Taylor & Francis.

Konz, S., 1992b, Vision of the workplace: Part II, *International Journal of Industrial Ergonomics*, **10**, 139–160.

Kroemer, K.H.E., 1989, Engineering anthropometry, *Ergonomics*, **32**, 767–784.

Kroemer, K.H.E., Snook, S.H., Meadows, S.K. and Deutsch, S., 1988, *Ergonomic Models of Anthropometry, Human Biomechanics, and Operator-Equipment Interfaces*, Washington, DC: National Academy Press.

Kroemer, K., Kroemer, H. and Kroemer-Elbert, K., 1994, *Ergonomics. How to Design for Ease and Efficiency*, Englewood Cliffs, NJ: Prentice-Hall.

Krohn, G.S., Sanders, M.S. and Peay, J., 1984, Workvests (lifevests) used for dredge mining, *Proceedings of Human Factors Society Annual Meeting*, pp. 478–482, Santa Monica, CA: The Human Factors and Ergonomics Society.

Krohn, R. and Konz, S., 1992, Best hammer handles, *Proceedings of the Human Factors Society*, pp. 413–417, Santa Monica, CA: The Human Factors and Ergonomics Society.

Kryter, K.D., 1985, *The Effects of Noise on Man*, 2nd edn, New York, Academic Press.

Knutsson, A., Åkerstedt, T. and Orth-Gorner, K., 1986, Increased risk of ischemic heart disease in shift workers, *Lancet*, **12**, 89–92.

Lawrence, J.S., 1955, Rheumatism in coal miners, Part II, *British Journal of Industrial Medicine*, **12**, 249–261.

Lehto, M.R., 1992, Design warning signs and warning labels: Part II – Scientific basis for initial guidelines, *International Journal of Industrial Ergonomics*, **10**, 115–138.

Lehto, M.R. and Miller, S.M., 1986, *Warnings*, Vol. 1, *Fundamentals, Design, and Evaluation Methodologies*, Ann Arbor, MI: Fuller Technical.

Lindh, M., 1980, Biomechanics of the lumbar spine, in Frankel, V.H. and Nordin, M. (Eds), *Basic Biomechanics of the Skeletal System*, Philadelphia, PA: Lea & Febiger.

Locke, E.A., 1983, The nature and causes of job satisfaction, in Dunnette, M.D. (Ed.), *Handbook of Industrial and Organizational Psychology*, New York: Wiley.

Loeb, M., 1986, *Noise and Human Efficiency*, Chichester: Wiley.

Loewenthal, A. and Riley, M.W., 1980, The effectiveness of warning labels, in *Proceedings of the 24th Annual Meeting of the Human Factors Society*, pp. 389–391, Santa Monica, CA: The Human Factors and Ergonomics Society.

Luczak, H., 1993, *Arbeitswissenschaft*, Berlin: Springer-Verlag.

Mackie, R.R., O'Hanlon, J.F. and McCauley, M.E., 1974, *A Study of Heat, Noise, and Vibration in Relation to Driver Performance and Physiological States (Technical Report 1735)*, Goleta, CA: Human Factors Research Inc.

Magora, A., 1974, Investigation of the relation between low back pain and occupation, *Scandinavian Journal of Rehabilitation Medicine*, **6**, 81–88.

Margolis, W. and Kraus, S.F., 1987, The prevalence of carpal tunnel syndrome symptoms in female supermarket checkers, *Journal of Occupational Medicine*, **29**, 953–959.

Maxion, R., 1984, Artificial intelligence approaches to monitoring systems integrity, in *Artificial Intelligence in Maintenance*, pp. 257–273, Brooks AFB, TX: Air Force Human Resources Laboratory.

McConville, J.T., Robinette, K.M. and Churchill, T., 1981, *An Anthropometric Data Base for Commercial Design Applications*, Yellow Springs, OH: Anthropology Research Project.

McGill, S.M. and Norman, R.W., 1986, Dynamically and statistically determined low back movement during lifting, *Journal of Biomechanics*, **18**, 877–885.

McGrath, J.J., 1976, *Driver Expectancy and Performance in Locating Automotive Controls (SAE Report SP-407)*, Santa Barbara, CA: Anacapa Sciences, Inc.

Meister, D., 1971, *Human Factors: Theory and Practice*, New York: Wiley.

Mellor, E.F., 1986, Shift work and flextime: how prevalent are they? *Monthly Labor Review*, **109**, 14–21.

Michael, E.D., Hutton, K.E. and Horvath, S.M., 1961, Cardiorespiratory responses during prolonged exercise, *Journal of Applied Physiology*, **16**, 997–1000.

Michel, D.P. and Helander, M.G., 1994, Effect of two types of chairs on stature change and comfort for individuals with healthy and herniated discs, *Ergonomics*, **37**, 1231–1244.

Miller, G.A., 1956, The magical number seven plus or minus two: some limits on our capacity for processing information, *Psychological Review*, **63**, 81–97.

Miller, J.C., 1992, *Fundamentals of Shift Work Scheduling*, Lakeside, CA: Evaluation Systems, Inc.

Miller, J.C. and Horvath, S.M., 1981, Work physiology, in Helander, M.G. (Ed.), *Human Factors/Ergonomics For Building and Construction*, New York: Wiley.

Mital, A. (Eds), 1991, Economics of flexible assembly automation: Influence of production and market factors, in Parsaei, H.R. and Mital, A. (Eds), *Economic Aspects of Advanced Production and Manufacturing Systems*, London: Chapman & Hall.

Mital, A., Nicholson, A.S. and Ayoub, M.M., 1993, *A Guide to Manual Materials Handling*, London: Taylor & Francis.

Monk, T.H., 1986, Advantages and disadvantages of rapidly rotating shift schedules – a circadian viewpoint, *Human Factors*, **28**, 553–557.

Monk, T.M. and Folkard, S., 1992, *Making Shift Work Tolerable*, London: Taylor & Francis.

Montemerlo, M.D. and Eddower, E., 1978, The judgemental nature of task analysis, in *Proceedings of the 22nd Annual Meeting of the Human Factors Society*, pp. 247–250, Santa Monica, CA: The Human Factors and Ergonomics Society.

Morris, N.M. and Rouse, W.B., 1984, *Review and Evaluation of Empirical Research in Troubleshooting (Report 8402-1)*, Norcross, GA: Search Technology Inc.

Nagamachi, M. and Yamada, Y., 1992, Design for manufacturability through participatory ergonomics, in Helander, M. and Nagamachi, M. (Eds), *Design for Manufacturability*, pp. 219–229, London: Taylor & Francis.

NASA, 1978, *Anthropometry Source Book, Volume II: A Handbook of Anthropometric Data*, Houston, TX: NASA.

National Institute for Occupational Safety and Health, 1989, NIOSH criteria for a recommended standard: occupational exposure to hand-arm vibration, *Report DHHS-NIOSH Publ. 89-106*, Cincinnati, OH: NIOSH.

National Institute for Occupational Safety and Health, 1992, *Health Hazard Evaluation Report, HETA 89-299-2230*, Cincinnati, OH: NIOSH.

National Research Council, 1983, *Video Displays, Work, and Vision*, Washington, DC: National Academy Press.

Nichols, D.L., 1976, Mishap analysis, in Ferry, T.S. and Weaver, D.A. (Eds), *Directions in Safety*, Springfield, IL: Charles C. Thomas.

Öquist, O., 1970, Kartläggning and Individuella Dygnsrytmer, Thesis, Department of Psychology, Göteborg University, Göteborg, Sweden.

Östberg, O., 1980, Accommodation and visual fatigue in display work, in Grandjean, E. and Vigliani, E. (Eds), *Ergonomic Aspects of Visual Display Terminals*, London: Taylor & Francis.

Parrish, R.N., Gates, J.L., Munzer, S.J., Grimma, P.R. and Smith, L.T., 1982, *Development of Design Guidelines and Criteria for User/Operator Transactions with Battlefield Automated Systems, Phase II Final Report: Volume II*, Alexandria, VA: US Army Research Institute for the Behavioral and Social Sciences.

Patrick, J., 1992, *Training: Research and Practice*, San Diego, CA: Academic Press.

Pheasant, S.T., 1986, *Bodyspace, Anthropometry, Ergonomics and Design*, London: Taylor & Francis.

Pheasant, S.T. and Stubbs, D., 1992, *Lifting and Handling. An Ergonomic Approach*, London: National Back Pain Association.

Poulton, E., 1978, A new look at the effects of noise: a rejoinder, *Psychological Bulletin*, **85**, 1068–1079.

Poulton, E.C., 1979, *The Environment at Work*, Springfield, IL: Charles C. Thomas.

Prabhu, G.V., Helander, M.G. and Shalin, V., 1992, Cognitive implications product structure on manual assembly performance, in Brödner, P. and Karwowski, W. (Eds), *Ergonomics of Hybrid Automation III*, pp. 259–272, Amsterdam: Elsevier.

Purswell, J.L., Krenek, R.F. and Dorris, A., 1987, Warning effectiveness: what we need to know, in *Proceedings of the Human Factors Society 31st Annual Meeting*, pp. 1116–1120, Santa Monica, CA: The Human Factors and Ergonomics Society.

Putz-Anderson, V. (Ed.), 1988, *Cumulative Trauma Disorders. A Manual for Musculoskeletal Diseases of the Upper Limbs*, London: Taylor & Francis.

Putz-Anderson, V. and Waters, T.R., 1991, 'Revision in NIOSH Guide to Manual Lifting', presentation at the Conference on a National Strategy for Occupational Musculoskeletal Injury Prevention – Implementation Issues and Research Needs, Cincinnati, OH: National Institute of Occupational Safety and Health.

Rahimi, M. and Hancock, P.A., 1988, Sensor integration, in N. Dorf (Ed.), *International Encyclopedia of Robotics*, New York: Wiley.

Rahimi, M. and Karwowski, W. (Eds), 1993, *Human–Robot Interaction*, London: Taylor & Francis.

Ramazzini, B., 1940 (1717), Wright, W. (Trans.) *The Disease of Workers*, Chicago, IL: University of Chicago Press.

Rasmussen, J., 1986, *Information Processing and Human Machine Interaction: An Approach to Cognitive Engineering*, Amsterdam: North Holland.

Ray, R.D. and Ray, W.D., 1979, An analysis of domestic cooker control design, *Ergonomics*, **22**, 1243–1248.

Robotic Industries Association, 1987, *American National Standard R15.06 for Industrial Robots and Robot Systems Safety Requirements*, Ann Arbor, MI: Robotic Industries Association.

Robotic Industries Association, 1989, *Proposed American National Standard of Human Engineering Design Criteria for Hand Held Control Pendants*, Ann Arbor, MI: Author.

Roebuck, J.A., Kroemer, K.H.E. and Thomson, W.G., 1975, *Engineering Anthropometry Methods*, New York: Wiley.

Rohmert, W. and Luczak, H., 1978, Ergonomics in the design and evaluation of a system for postal video letter coding, *Applied Ergonomics*, **9**, 85–95.

Rosenbrock, H.N., 1983, Seeking an appropriate technology, in *Proceedings of IFAC Symposium on Systems Approach to Appropriate Technology Transfer*, Vienna: IFAC.

Rutenfranz, J., Haider, M. and Koller, M., 1985, Occupational health measures for nightworkers and shiftworkers, in Folkard, S. and Monk, T.H. (Eds), *Hours of Work – Temporal Factors in Work Scheduling*, pp. 199–210, New York, Wiley.

Safir, A. (Ed.), 1980, *Refraction and Clinical Optics*, Hagerstown, PA: Harper & Row.

Salmoni, A.W., Schmidt, R.A. and Waller, C.B., 1984, Knowledge of results and motor learning: a review and critical appraisal, *Psychological Bulletin*, **95**, 355–386.

Sanders, M.A., 1980, Personal communication, Canyon Research Group, Westlake Village, CA, USA.

Sanders, M.S. and McCormick, E.J., 1993, *Human Factors in Engineering and Design*, New York: McGraw-Hill.

Sauter, S.L., Chapman, L.S. and Knutson, S.J., 1985, *Improving VDT Work*, Madison, WI: Department of Preventive Medicine, University of Wisconsin.

Scherrer, J., 1981, Man's work and circadian rhythm through the ages, in Reinberg, A., Vieux, N. and Andlauer, P. (Eds), *Night and Shift Work: Biological and Social Aspects*, pp. 1–10, Oxford: Pergamon Press.

Schoenmarklin, R. and Marras, W., 1989, Effect of hand angle and work orientation on hammering: II. Muscle fatigue and subjective ratings of body discomfort, *Human Factors*, **31**, 413–420.

Scholey, M. and Hair, M., 1989, Back pain in physiotherapists involved in back care education, *Ergonomics*, **32**, 179–190.

Sen, R.S., 1989, Personal communication, SUNY at Buffalo, Buffalo, NY, USA.

Shalin, L., Prabhu, G.V. and Helander, M.G., 1995, A Cognitive Perspective on Manual Assembly, *Ergonomics*, in press.

Shurtleff, D.A., 1980, *How to Make Displays Legible*, LaMirada, CA: Human Interface Design.

Shute, S.J. and Starr, S.J., 1984, Effects of adjustable furniture on VDT users, *Human Factors*, **26**, 157–170.

Silverstein, B.A., Fine, L.J. and Armstrong, T.J., 1987, Occupational factors and carpal tunnel syndrome, *American Journal of Industrial Medicine*, **11**, 343–358.

Simon, H.A., 1969, *The Science of the Artificial*, Cambridge, MA: MIT Press.

Simon, H.A., 1974, How big is a chunk?, *Science*, **183**, 482–488.

Singleton, W.T., 1962, *Ergonomics for Industry*, London: Department of Scientific and Industrial Research.

Snyder, H.L., 1988, Image quality, in Helander, M.G. (Ed.), *Handbook of Human–Computer Interaction*, Amsterdam: North Holland.

Sperry, W., 1978, Aircraft and airport noise, in Lipscomb, D. and Taylor, A. (Eds), *Noise Control: Handbook of Principles and Practices*, New York: Van Nostrand Reinhold.

Swain, A.D. and Guttman, H.E., 1980, *Handbook of Human Reliability Analysis with Emphasis on Nuclear Power Plant Applications (NUREG/CR-1278)*, Washington, DC: US Nuclear Regulatory Commission.

Swedish Work Environment Fund, 1985, *Making the Job Easier: An Idea Book*, Stockholm: SWEF.

Tasto, D.L. and Colligan, M.J., 1977, *Shift Work Practices in the United States*, Cincinnati, OH: National Institute for Occupational Safety and Health.

Tepas, D.I. and Monk, T.H., 1986, Work schedules, in Salvendy, G. (Ed.), *Handbook of Human Factors*, New York: Wiley.

T.G. and R.L., 1975, Conveyor belt sickness, *National Safety News*, **117**, 37.

Tichauer, E.R., 1966, Some aspects of stress on forearm and hand in industry, *Journal of Occupational Medicine*, **8**, 63–71.

US Department of Defense, 1989, *Military Standard 1472D*, Washington, DC: USDD.

US Department of Labor, 1980, *Noise Control. A Guide for Workers and Employers*, Washington, DC: USDOL/OSHA.

US Department of Labor, 1982, *Back Injuries Associated with Lifting (Bulletin No. 2144)*, Washington, DC: Bureau of Labor Statistics.

Van Cott, H.P. and Kinkade, R.G., 1972, *Human Engineering Guide to Equipment Design*, Washington, DC: US Government Printing Office.

Van Wely, P., 1970, Design and disease, *Applied Ergonomics*, **1**, 262–269.

Vora, P.R., Helander, M.G. and Shalin, V.L., 1994,Evaluating the influence of interface styles and multiple access paths in hypertext, *Proceedings of CHI'94*, pp. 323–329, New York: ACM.

Ward, W.D., 1976, Transient changes in hearing, in *Proceedings of the International Congress on Man and Noise*, pp. 111–122, Turin: Edizioni Minerva Medica.

Waters, T.R., Putz-Anderson, V., Garg, A. and Fine, L.J., 1993, Revised NIOSH regulation for the design and evaluation of manual tasks, *Ergonomics*, **36**, 749–776.

Webb, R.D.G., 1982, *Industrial Ergonomics*, Toronto: Industrial Accident Prevention Association.

Webster, B.S. and Snook, S.H., 1994a, The cost of comparable upper extremity cumulative trauma disorder, *Journal of Occupational Medicine*, in press.

Webster, B.S. and Snook, S.H., 1994b, The cost of 1989 worker's compensation low back pain claims, *Spine*, in press.

Webster, J., 1969, *Effects of Noise on Speech Intelligibility (ASHA Report 4)*, Washington, DC: American Speech and Hearing Association.

Weinstein, N., 1977, Noise and intellectual performance: a confirmation and extension, *Journal of Applied Psychology*, **62**, 104–107.

Welford, A.T., 1968, *Fundamentals of Skills*, London: University Paperback.

White, B. and Samuelson, P., 1990, Repetitive motion trauma in automotive parts manufacturing, in Karwowski, W. and Rahimi, M. (Eds), *Ergonomics of Hybrid Automated Systems*, Vol. 2, Amsterdam: Elsevier.

Whitefield, A., 1986, Human factors aspects of pointing as an input technique in interactive computer systems, *Applied Ergonomics*, **17**, 97–104.

Wickens, C.D., 1992, *Engineering Psychology and Human Performance*, New York: Harper Collins.

Wiener, E.L. and Nagel, D.C. (Eds), 1988, *Human Factors in Aviation*, San Diego, CA: Academic Press.

Winkel, J., 1990, Personal communication, Buffalo, NY, USA.

Winkel, J. and Oxenburgh, M., 1993, Towards optimizing physical activity in VDT/office work, in Sauter, S.L., Dainoff, M.J. and Smith, M.J. (Eds), *Promoting Health and Productivity in the Computerized Office: Models of Successful Ergonomic Intervention*, Amsterdam: North Holland.

Wogalter, M.S., Desaulniers, D.R. and Godfrey, S.S., 1985, Perceived effectiveness of environmental warnings, in *Proceedings of the 29th Annual Meeting of the Human Factors Society*, pp. 664–669, Santa Monica, CA: The Human Factors and Ergonomics Society.

Woodson, W.E., 1981, *Human Factors Design Handbook*, New York: McGraw-Hill.

Woodson, W.E. and Conover, D.W., 1964, *Human Engineering Guide to Equipment Design*, Berkeley, CA: University of California Press.

Wotton, E., 1986, Lighting the electronic office, in Lueder, R. (Ed.), *The Ergonomics Pay Off, Designing the Electronic Office*, Toronto: Holt, Rhinehart & Winston.

Wright, G. and Rea, M., 1984, Age, a human factor in lighting, in *Proceedings of the 1984 International Conference on Occupational Ergonomics*, pp. 508–512, Rexdale, Ontario: Human Factors Association of Canada.

Yerkes, R.M. and Dodson, J.D., 1908, The relation of strength of stimulus to rapidity of habit-formation, *Journal of Comparative Neurology of Psychology*, **18**, 459–482.

Zandin, K.B., 1990, *MOST Work Measurement System*, New York: Marcel Dekker.

Zenz, C., 1981, Physical Health Hazards in Construction, in Helander, M.G. (Ed.), *Human Factors/Ergonomics for Building and Construction*, New York: Wiley.

Zimolong, B., 1985, Hazard perception and risk estimation in accident causation, in Eberts, R.E. and Eberts, C.G. (Eds), *Trends in Ergonomics/Human Factors*, Vol. 2, pp. 463–470, Amsterdam: North Holland.

Zipp, von P., Haider, E., Halpern, N., Mainzer, J. and Rohmert, W., 1981, Untersuchung zur ergonomischen Gestaltung von Tastaturen, *Zentralblatt für Arbeitsmedizin, Arbeitsschutz, Prophylaxe und Ergonomi*, **31**, 326–330.

Zwaga, H. and Easterby, R., 1982, Developing effective symbols for public information: the ISO testing procedure, in *Proceedings of 1982 Congress of the International Ergonomics Association*, pp. 512–513, Santa Monica, CA: The Human Factors and Ergonomics Society.

Appendix: The Use of an Ergonomics Checklist in Manufacturing

There are two major uses of checklists:

1. As a memory aid during inspection of industrial plants.
2. As a tool for systematic data collection.

In the case of a memory aid, we assume that during a plant inspection it may be difficult to think of and remember important design details. Thereby, by using a checklist, it is possible to systematically cover all the important ergonomic issues. However, there is a danger in using checklists. Just as in the case of task analysis, there is no fixed method. The types of item included in the checklist depend on the application. One would devise a very different checklist for automated manufacturing than for manually oriented manufacturing. Different items would be checked in a process plant, as compared with a small workshop. It is therefore necessary to develop a checklist that is suitable for the purpose. One should not uncritically accept the checklist supplied below. It needs to be complemented to cover the items that are of particular importance for the environment being investigated. It may be a matter of negotiation between management workers and labour unions to decide which items should be included. It is also possible that a checklist could be developed in a 'quality circle' or an 'ergonomic task group'. Furthermore, checklists can have an emphasis on or bias towards different issues such as environmental hygiene, safety and injuries, productivity, or operator comfort and convenience. Thus, the items on the checklist will depend on the criteria under evaluation.

The other application is to use a checklist as a tool for the systematic investigation of workplaces. Imagine, for example, that there are 500 microscope workstations in a plant. It may then be of interest to collect statistics on the ergonomic design features associated with these workplaces. Issues of adjustability, work posture, and illumination may be particularly important. In this case, the statistics gathered can serve as a management tool for evaluating how many workstations need refurbishing and how many are in good condition. The checklist can also be used to produce a list of priorities for upgrading workstations. It would then be possible to predict the cost of certain types of upgrade and to compare this with the expected benefits.

Checklists can be made more detailed. Instead of a checkmark there could be an evaluation on a scale of, say, 1–5, or the checklist could be complemented with a questionnaire so that workers themselves do the checking.

A somewhat less ambitious approach is the use of a 'survey checklist' (Eastman Kodak Co., 1983). This type of a checklist is not as complete as many others that have been developed for ergonomic surveys (Van Wely, 1970; Woodson, 1981; International Labor Office, 1987). Rather, the survey checklist is more generic and problem oriented. The intention is that the list of items could lead to further discussion and a more detailed analysis.

The checklist presented below is problem oriented and intended to inspire design improvements. It can be used as a basis for discussions at work in groups or between individuals. This can lead to many innovations and improvements of both the task and the workplace. In some cases it may not be possible to justify the cost for ergonomic improvements. Such instances can also be documented in the responses to the questions in the checklist.

ERGONOMIC CHECKLIST TO ENHANCE PERFORMANCE, SAFETY, AND COMFORT.

A. Physical demands	Yes	If No, then Why?	How to redesign?
Are hands at a convenient working height for the task?			
Are the joints mostly in a convenient neutral position?			
Are the wrists mostly in a straight, neutral posture?			
Can operator assume several different postures while working?			
Is this a dynamic rather than a static task?			
Can the task be performed with the torso and the head facing forward?			
Are primary items located within easy reach?			
Is frequent lifting below 20 kg (45 lb)?			
Is occasional heavy lifting less then 25 kg (55lb)?			
Are items to be lifted positioned between knuckle and shoulder height?			
Are there convenient aids for manual materials handling?			

Are there handles on items which are otherwise difficult to lift?

Are handtools appropriate for the task?

Are handtools comfortable and safe to use?

For sitting tasks:

Are the feet firmly supported on the floor or by using a footrest?

Can the backrest be utilized while performing the task?

Are the elbow joints mostly at an intermediate angle?

Are primary items located within easy (5th percentile) reach - about 40 cm.

Is head bent slightly forward - rather than backward ?

B. Task visibility	**Yes**	**If No, then Why?**	**How to redesign?**
Are displays and dials easy to see from normal work position?			
Is printed or displayed text large enough for reading about 18-25 min. of arc?			
Are eyeglasses appropriate for task viewing distance?			
Is illumination level uniform throughout working area?			
Are illumination levels appropriate? About 500 lux for VDT work About 1000 lux for coarse assembly About 2000 lux for fine assembly			
Is direct glare from illumination sources and windows avoided?			
Is indirect (reflected) glare avoided?			
Is luminance contrast ratio in immediate task area less than 20:1?			

C. Mental demands	**Yes**	**If No, then Why?**	**How to redesign?**
Does the task involve moderate short-term memory load - rather than high?			
Does the task involve few simultaneous factors - rather than several?			
Is operator performance unpaced - rather than paced by the task?			
Is the task varying - rather than repetitive and monotonous?			
Can operator errors and slips easily be corrected?			
Are special memory aids used?			
Do displays and controls follow population stereotypes?			
Is the task easy to learn - rather than difficult?			

D. Machine Design	**Yes**	**If No, then Why?**	**How to redesign?**
Are tasks appropriately allocated between operators and machines?			
Are manual controls easy to reach?			
Are manual controls easy to distinguish from each other?			
Are all machine functions and displays visible to the operator?			
Can machine functions be handled through one command/control?			
Are all controls on the machine necessary for the job?			
Are location of controls and tools the same for similar machines?			
Are memory aids used as a reminder			

of difficult task information?

Is it possible to operate machine without bending, twisting and far reaching?

Is there adequate body clearance for handling and maintenance tasks?

Are machine symbols and icons readily understood?

Are labels used to inform and remind operators of task information?

Are labels/symbols used to designate locations for frequently used items?

E. VDT Tasks	**Yes**	**If No, then Why?**	**How to redesign?**
Are screens positioned perpendicular to windows?			
Can reflected glare on the screen be avoided?			
Is the display located below a horizontal plane through the eyes?			
Do the locations of display, documents and keyboard make it possible to sit straight without twisting the body?			
Is a QWERTY keyboard used?			
Are software functions understood and easy to use?			
Are software functions and computer task routines easy to access?			

F. Safety	**Yes**	**If No, then Why?**	**How to redesign?**
Are there appropriate warning signs as a reminder of task hazards?			
Is wording on warning signs relevant and informative?			
Are warning signs positioned where operators look?			

Is the workplace organized and clean with excellent house keeping?

Are the floors even without drains or pit marks?

Is it possible to perform the task without safety glasses or protective clothing?

Has company established safety procedures and rules ?

Are safety rules and procedures prioritized by management and enforced?

Does company analyze each reported accident or injury to improve safety?

Do newly hired workers receive safety training?

Do safety training programs present relevant task specific information?

Are potential hazards clearly visible from the operators position?

Have machine safety devices been installed, e.g. lockouts, and guards?

G. Ambient Environment	**Yes**	**If No, then Why?**	**How to redesign?**

Is ambient noise below 85 dBA to protect against hearing damage?

Is ambient noise level below 55 dBA to facilitate verbal communication?

Is there a program to reduce noise pollution by redesign of machines and the work environment?

Are vibration levels and frequencies so low as to not affect job performance?

Is the temperature and humidity within a comfortable range?

Is it possible to perform work tasks without protective equipment?

Can all work tasks be performed

without risk of electric shock?

H. Product and Process Design	**Yes**	**If No, then Why.?**	**How to redesign?**
Has product design been modified to improve productivity?			
Has product design been modified to create better jobs?			
Have the best machines been selected that maximize productivity?			
Have the best machines been selected that maximize operator convenience?			
Have processes been located so as to improve productivity?			
Have processes been located to improve operator convenience?			
Have machines and processes been selected to optimize task allocation between operators and machines?			

Index

accelerator reference point (ARP) 24, 25
accessibility, design for 174
acclimation 33
acclimatization 3–4
accommodation 83–5
acromion height 23
active processing of warning signs 118–19
adenosine triphosphate (ATP) 29
aerobic metabolism 32
affirmative statements 117
age
  and eyesight 83–6
  and hearing 134
  and work capacity 30, 31
aeroplanes 24, 25
anaerobic metabolism 31–2
annoyance, noise 139–42
anthropometrics design motto 20
anthropometry 17
  definition of measures 21–4
  design procedure 25–8
  for industrial design 24–5
  measuring human dimensions 17–21
arm rests 93–4
assembly, design for 157–62
  automation, desire for 155–7
  Boothroyd's principles 163
  human factors principles 165–9
  motion time measurement analysis 163–5
asymmetric parts 160–1
auditory feedback 166
Australia 69
automatic test equipment (ATE) 171–2
automation
  Boothroyd's principles 163
  desire for 155–7
  printer assembly case study 12, 14–15

back injuries 4, 5, 69–70
  from lifting 39–41
barriers, physical 177–8
base parts 157
basic metabolic rate (BMR) 30
bending 40
Bennett's bend 77
bent back-straight knees lifting technique 44
bifocal lenses 95
bin-assembly compatibility 107–8
biomechanical model for lifting 41–3
body temperature 33
Boothroyd's principles 163
Broadbent, D. 138–9
buddy system 13
buttock-knees depth 23
buttock-popliteal depth 23, 26

calculator button layout 102–3
cameras 179
capacitive sensors 179
card assembly case study 7–12
cardiovascular disease, shift workers 146
career path 13
carousels 54
carpal tunnel syndrome 5, 65–7, 70
  hand tools 76
cars 24–5
Carter, Jimmy 91, 92
carts 54
cataracts 95
chains 178
chairs
  anthropometry 20, 21, 26–7
  card assembly case study 9
  printer assembly case study 13
  VDT workstations 92–3
checklists 197–9
circadian rhythms 144–5
clothes wringing disease 68
clothing
  anthropometry 23
  card assembly case study 9, 10
clouding of vision 86
coding 108–11, 160, 172–3
colour coding 108–9, 111, 160, 172–3
colour rendering index (CRI) 88–9
combined parts 158
comfort climate 36–7
communication
  card assembly case study 9, 10
  noise interference 140–2
computer-aided design 24
computers 1–2
  input devices 103–4
  see also visual display terminals
conductive hearing loss 133, 134
construction workers
  physical workload 29
  symbols, standardization of 115–16
continental rota 149
continuous flow manufacturing 10
contrast ratio 80, 81, 82, 99
contrast, measurement of 80–1
controls
  checklist 199

coding 108–11
computer input devices 103–4
control-response compatibility 106–8
emergency 111
manual, appropriateness 101–3
movement stereotypes 104–6
organization 111–13
printer assembly case study 14
convection 34
conversion tables 9
'conveyor sickness' 64
conveyors 54
posture 63–4
cooker controls 106–7
'correct lifting technique' 43–6
cost efficiency of improvements
card assembly case study 10–12
illumination 87–9
cubital tunnel syndrome 67
cumulative trauma disorder (CTD) 4–5
causes 69–71, 159
hand tools 75–6
minimization 70, 71
types 65–8
cycle time 12–13

Dart's disease 78
data collection, checklists for 197
database, anthropometric 25
daylight 145
De Quervain's disease 68
Department of Defense (USA) 3, 21
Department of Defense (USSR) 3–4
Department of the Interior 3
design 1–2
for automation (DFA) 12, 155
for human assembly (DHA) 157, 165–6, 168
for manufacturability (DFM) 155
diagnostic tools 171–2
dials 199
diffuse reflection 86, 87
direct glare 86
discomfort glare 80–1
disks, spinal 41–42
displays
checklist 199
printer assembly case study 14
doctors 4
dosimeters 134
drills 10, 11
drivers, whole body vibration 151–2
durable materials 161–3

ear muffs 134
ear plugs 134
Eastman Kodak 3
elbow height 23
elbow rests 94
elbow-to-elbow breadth 23
electromagnetic radiation 95
emergency controls 111
emergency situation 46
energy expenditure 29–30
engineers 4
environment 198
epileptics, and shift work 150
Ergonomics Research Society 4
error codes, job aids 127
etching, VDT screens 97
European Community lifting guidelines 50, 51
evaporated heat loss 34
external precision grip 73, 74
eye, ageing process 83–6
eye reference point (ERP) 24, 25
eyeglasses, VDT operators 95

fasteners 158–9
fatigue
due to physical workload 33
shift workers 147–8
fault identification, ease of 171–3
Federal Aviation Administration 3
Federal Highway Administration 3
feedback
auditory 166
card assembly case study 9
printer assembly case study 13
tactile 166
training 124, 126
visual 166
feeders 161, 164
fences 177
filters, VDT screens 97–9
Fitts' law 161
fluorescent light 89
footrests
anthropometry 20–1
VDT workstations 93
France 5
frosting, VDT screens 97
functional forward reach 23
functional overhead reach 23

gastrointestinal problems, and shift work 150
gender differences, human dimensions 17, 20, 21, 74
General Motors 155
genetic differences, human dimensions 17
Germany
labour unions 4, 149
shift work 148, 149
VDT workstations 91
Gilbreth, Frank 4
Gilbreth, Lillian 4
glare 86
VDT screens 96–9

gloves 10
graphics tablets 104
gravity feed conveyors 54
grey films 96
grip breadth, inside diameter 23
grip strength 74, 75

hacksaws 76–7
hammers 77
hand height 61–3
hand reference point (HARP) 24, 25
hand size 74
hand tools 73–8
  anthropometry 21
  coding 110-11
  printer assembly case study 14
handtrucks 53
handing suspended tools 53
head-up displays 25
Health and Safety Commission 50, 52
hearing
  loss 133–4
  protectors 134
heart disease, shift workers 146
heart rate 30, 33
heat exposure 34
heat stress 33–7
heating 14
hip breadth 23
hip reference point (HRP) 24–5
horizontal louvres 96
horizontal transportation 52–4
housekeeping 9
human dimensions 17–21
human errors 121
Human Factor Society 4
human factors engineering 1
humidity 34

IBM Corporation 3
  anthropometry 19–20
  card assembly case study 7–12
  paper picking mechanism 155–6
  printer assembly case study 12–15, 157
  robot safety 180
Illuminating Engineering Society (IES) 82–3, 84
illumination
  cost efficiency 87–9
  measurement 79–80
  photometers 81–2
  recommended levels 82–3
  see also lighting; lumination
image processing 179
incandescent light 89
India 58–9
indirect glare 86
indirect lighting 86–7, 88
information overload 118
infrared sensors 179
inspection
  lighting for 89–90
  shift work 147–8
insurance premiums 4
integrated parts 158
interdisciplinary nature of ergonomics 2–3, 4
internal precision group 73, 74
International Ergonomics Association 4
interpupillary distance 23
ionizing radiation 95
irritation, noise 139–42

Jastrzebowski, Wojciech 1
job aids 127–8
job rotation
  card asembly case study 9
  printer assembly case study 13
job satisfaction, design for 168–9
joysticks 104
just in time (JIT) manufacturing 52, 54

keyboards, low-profile 91–2
knowledge-based tasks 23
  noise levels 137–8
  shift work 147
knuckle height 22–3
Kodak 172
Korean War 3

labels
  coding by 110
  design of 115, 117
  for maintainability 172
labour unions 4, 149
lactic acid 31–2
lathes 17, 18
left-handed people 74–5, 112
leisure activities 30
levelators 53
Lexmark 180
lifting
  equipment 40–1
  manual 39
    back injury statistics 39–41
    biomechanical model 41–3
    'correct lifting technique' 43–6
    guidelines and standards 46–50
    materials handling aids 50–4
lifting index (LI) 50
light pens 103, 104
lighting
  card assembly case study 8
  indirect 86–7
  printer assembly case study 14
  special-purpose 89–90
  VDT workstations 96
  see also illumination; lumination

loading of conveyors 64
location
  coding by 108
  of warning signs 118
louvres 96
low-profile keyboards 91-2
lumbar support 93
lumination
  measurement 79-80
  photometers 81-2
  see also illumination; lighting

maintainability, design for 171-6
managers 17-18
manipulation, ease of 166, 174, 175
manufacturing
  anthropometric design 24, 25
  back injuries 39
materials handling
  aids 50-4
  card assembly case study 9
mean value 18-19
median nerve 65, 66
memory aids, checklists as 197
memory span 119-20
  and noise levels 138
mental load 198
mental models 168
metabolism
  physical workload 29-30, 35
  during work 30-2
metric to decimal conversion 9
metropolitan rota 149
micro-louvre filters 99
micro-mesh filters 99
mock-ups, full-scale 27
mode of operation, coding by 110
modulation contrast 80
monotonous jobs 9
morningness 150
motion time measurement (MTM) 163-5
mouse, computer 104
movement envelopes 112, 113
music 9

NASA 3
National Bureau of Standards 3
National Highway Traffic Safety Administration 3
National Institute of Occupational Safety and Health 3
neck pain 5
negative statements 117
negative transfer of training 106
Netherlands 5
neural hearing loss 133-4
NIOSH equation, lifting evaluation 46-50
noise levels
  analysis and reduction 134-6
  card assembly case study 10
  communication, interference with 139-42
  hearing loss 133-4
  performance, effects on 136-9
  printer assembly case study 14
normal distribution 18-19
Nuclear Regulatory Commission 3
nurses 4

octave-band analysis 135
overhead balancers 54
overhead cranes 53
  controls 104-6
oxygen debt 32
oxygen uptake 30, 33

packaging 49-50
'Panama Canal conveyor' 63
paper picking mechanism 155-6
part-task training 126
passive statements 117
percentiles 18-19, 26
perceptual load 198
personal items, storage 13
photoelectric beams 178
photometers 81-2
physical barriers 177-8
physical condition 30, 31
physical workload 29-33
  checklist 199
pliers 76
polarized lights 8
popliteal height 23, 26
postures
  choice 59-60
  at conveyors 6-4
  examples 55-8
  hand height and table height 60-3
  poor 4, 58-9
  VDT workstations 91-4
  and vision 85-6
  workstation design 112
Poulton, E. C. 138-9
power grip 73-4
power law of practice 128-30
precision grip 73-4
predetermined time-and-motion studies (PTMS) 112, 163
predictive mean vote (PMV) 36-7
preferred noise criterion (PNC) curves 140-1
preferred speech interference level (PSIL) 141-2
presbycusis 134
presbyopia 94
pressure-sensitive mats 178-9
printer assembly case study 7, 12-15, 157
productivity 5
  card assembly case study 10
  shift workers 146-7

Proprinter manufacturing line case study 12–15
protective clothing 10
psychologists 4
punch press stock loading 48–9

quality control
  lighting for 89–90
  and shift work 148
quality groups 3
quarter-wave cooking 98

radiated heat 34
radiation, screen 95
reflectance 79
reflected glare 86
  VDT screens 96–9
reflected lighting 86–7, 88
relevance of warning signs 118
repetitive motion injury *see* cumulative trauma disorder
reversed video 99
Reynaud's disease 77
right-handed tools 74–5
robots
  for assembly 155–7, 164
  safety 177–80
  see also automation
rollable platforms 53
rolling containers 53
rule-based tasks 122–3
  noise levels 137, 139
  shift work 147

sacrum 41
safety
  improvements 5
  machines and robots 177–80
Scandinavia
  labour unions 4, 149
  shift work 148, 149
  VDT operators 5, 70
schedules, shift work 148–9
scientific study of work 4
sea sickness 151
segmental vibration 77–8, 151
self-levelling tables 53, 54
self-locating parts 161, 162
shape coding 109–10, 111
shift work 143–4
  circadian rhythms 144–5
  improvements 147–50
  overlap 9
  performance and productivity, effects on 146–7
  problems 145–6
shoes 20, 23
short-term memory 119–20
  and noise levels 138
shoulder height 23
shoulder pain 5
shoulder tendonitis 68
sit-standing 59–60
sitting 59–60
  India 58–9
  VDT workstations 91–4
  see also chairs
sitting elbow height 23
sitting eye height 23
sitting height 23
size
  coding by 109
  of warning signs 118
skill-based tasks 122–3
  noise levels 137, 139
  shift work 147
snap and insert assembly 158
social life, shift workers 146, 147
Society of Automotive Engineers 115
sodium light 89
sonar 179
sound level meters 131, 134
spatial compatibility 166–8
spectacles, VDT operators 95
specular reflections 97
spine structure 41–2
spinning-jenny 4
spinning-mule 4
springs 159–60
stable materials 161–3
standard deviation 18–19
standing 59–60
stature 23
stomach problems, shift workers 146
straight back-bent knees lifting technique 43–4
stray illumination 86
stress
  heat 33–7
  and performance 138–9, 153
  shift workers 148
survey checklists 197–9
sweating 33, 34
Sweden, traffic signs 116–17
Switzerland, back injuries 5
symbols, design of 115–17
symmetrical parts 160

tables
  anthropometry 20–1, 26–7
  materials handling 53
  posture issues 60–3
tack boards 13
tactile feedback 166
task analysis 124–6
task design measures 12–13
task illumination 86, 87
Taylor, Frederick 4

telephone button layout 102–3
temporary myopia 94
temporary threshold shift (TTS) 133
tendonitis 68
tenosynovitis 5, 68
  hand tools 76
terminal glasses 95
testability, design for 173–4
therbligs 4
thermal balance equation 34
thermocomfort 36
thermoplasts 158
thermoregulation 33–4
thigh clearance 23
  VDT workstations 91
thin-film coating 98
third-octave-band analysis 135
thoracic outlet syndrome 68
Three Mile Island nuclear accident 147
tibial height 21–2
tilted VDT screens 97
time-and-motion study 4
tolerances
  to noise 139
  of part mating 161, 162
touch screens 103–4
track-balls 104
traffic signs 116–17
training
  content and methods 122–3
  development 123–6
  in ergonomics 10
  in manual materials handling 45–6
  manufacturing skills 126–8
  need for 121–2
  power law of practice 128–30
  transfer of 106, 121, 126, 168
transportation 50–4
trigger finger 68
troubleshooting, design for 173–4
twisting 40
typing 24, 25

ulnar nerve 67
ultrasound 179
Union of Soviet Socialist Republics 3–4
United Kingdom
  back injuries 39
  history of ergonomics 4
  lifting guidelines 50, 52
  musculoskeletal disease, costs 5
United States of America
  back injuries 39
  compensation premiums 4
  control movements stereotypes 104
  cumulative trauma disorder 65
  history of ergonomics 3, 4
  noise exposure 132
  shift workers 143, 149
  VDT operators 5, 70
unloading, from conveyors 64

vacuum lifts 53
ventilation 14
vertical louvres 96
vertical transportation 52, 54
vibration
  discomfort, sources of 151–4
  hand tools 77–8
  and noise 136, 153
vibratory bowl feeders 161
viewing angle, VDT screens 91
viewing distance 94
visibility 166
visual display terminals (VDTs)
  design 91
    posture 91–4
    radiation 95
    reflections and glare 96–9
    visual fatigue 94–5
  injuries 5, 70–1
  interdisciplinary knowledge 1
  luminance 81, 82
visual displays, design of 115–20
visual fatigue 94–5
visual feedback 166

warning signs 117–20
washers 159
welding curtains 178
wet bulb globe temperature (WBGT) 34–5, 36
white finger disease 77
whole body vibration 151–4
whole-task training 126
windows
  discomfort glare 80–1
  VDT glare 96–7
work groups 4
workplace characteristics 198
workstation ergonomics
  anthropometry 17, 24–8
  organization of items 111–13, 114
  postures 55–7
  printer assembly case study 13–14
wrist rests 94
written signs 117

X radiation 95
Xerox 172

Yerkes-Dodson's law 138, 153

Zeitgeber 145